Microplasma Sprayed
Hydroxyapatite Coatings

Microplasma Sprayed Hydroxyapatite Coatings

Arjun Dey
Anoop Kumar Mukhopadhyay

CRC Press
Taylor & Francis Group
Boca Raton London New York

CRC Press is an imprint of the
Taylor & Francis Group, an **informa** business

CRC Press
Taylor & Francis Group
6000 Broken Sound Parkway NW, Suite 300
Boca Raton, FL 33487-2742

First issued in paperback 2017

ISBN-13: 978-1-4822-5093-0 (hbk)
ISBN-13: 978-1-138-74886-6 (pbk)

Visit the Taylor & Francis Web site at
http://www.taylorandfrancis.com

and the CRC Press Web site at
http://www.crcpress.com

To

Manju, Brishti, Tutum,

and

Omprita

Contents

Preface

Beginning of the Journey

We started work on bioceramics, in particular the hydroxyapatite (HAp) coatings, almost a decade ago at the CSIR–Central Glass and Ceramic Research Institute at Kolkata in India. After a preliminary study we understood that this subject was not only restricted within the domain of biological science, but also linked to multiple disciplines, such as manufacturing technology, mechanical engineering, nanotechnology, and finally, materials science. Thus, in this book, *Microplasma Sprayed Hydroxyapatite Coatings*, a major focus has been dedicated to in-depth discussions from the materials science point of view, in addition to the conventional, biological investigation point of view. It needs to be explained why *hydroxyapatite coatings* and the materials science approach come into the picture. It is very well recognized today that there has been an enormous growth in the use of medical implants. The growth has been particularly phenomenal during the past couple of decades. You may wonder why this is so.

Well, there are a couple of very important pertinent reasons. This has primarily happened due to an almost commensurate enhancement in life expectancy, drastic change in accident frequency, along with use of high-speed transport systems, and, of course, to a major extent, the paradigm improvement or at least the demand for that in the biomedical implant materials themselves and the corresponding surgical technologies themselves. Even given this scenario, conservative statistics would reflect that millions of patients suffering from bone defect–related, oral, and maxillofacial disorders are indeed eagerly looking forward to a long-term solution to regain at least a nominal, if not the very desired, high quality of life. It is indeed very heartening to note that diseases and problems caused by damaged or diseased bone tissue today represent an annual market that now exceeds €50 billion worldwide. Moreover, the global demand for medical ceramics and ceramic components was about US$10 billion in 2010. The forecast is that it will have an annual average growth of about 6.5% to scale a height of about US$15 billion by 2015 that will eventually expand to about US$20 billion by 2018.

Now, let us consider a very interesting population statistics scenario. Globally, the number of older persons (aged 60 years or over) was 841 million people in 2013. It is now expected that this number will more than double and reach a population of about 2 billion by 2050. In fact, it has been projected that by 2047, for the very first time in human history, the number of children will be smaller than the amount of older people. What turns out to be of critical importance in this perspective is that, as of now, the developing countries possess about two thirds of the world's older persons, who represent the most prospective population exposed to the risk of bone fracture, and hence deeply in need of bioactive ceramic coatings for repair of bone defects. A fact of the additional and paramount importance in this context is that unfortunately the size of the older population in less developed and developing regions is growing faster than those in the more developed regions. As a matter of a big healthcare problem, it now emerges that by 2050 every 8 out of 10 persons belonging to the world's older population are predicted to be in the less developed and developing regions, e.g., many Southeast Asian, African, and Latin American countries. In other words, the issue is of extraordinary socioscientific, socioeconomic, and humanitarian scientific concerns as well. This book is actually intended to be a humble milestone in this overall global scenario.

We started extensive work in this field when we realized that, especially in the case of hip joint replacement, loosening of a metallic prosthesis fixed with polymethylmethylacrylate (PMMA) bone cement causes the patient to undergo painstaking revision surgery, which is a major problem not only for the patient and the surgeon, but also the biomedical technology itself. In fact, the global recognition of this problem led to the development of cementless fixation through the novel introduction of a bioactive hydroxyapatite (HAp) coating on biomedical-grade metallic implants. It was a natural follow-up action of this development that a wide variety of different coating methods have evolved to make the HAp coatings on metallic implants more reliable. However, among all such developmental efforts, the atmospheric plasma spraying (APS) or macroplasma spraying (MAPS) method has ultimately been accepted commercially. Nonetheless, even this most welcome development did not resolve the issue, and the challenge of developing more reliable HAp coatings was far from resolved. A lot of questions eventually surfaced, for instance: Are the coatings adherent enough? Do they have adequate bonding with the metallic implant they coat? Do they possess adequate stability in and outside of the physiological environment? How good is their biofunctionality, e.g., in both in vitro and in vivo atmospheres, etc.? Moreover, it has now been realized that the conventional high-power plasma spray (i.e., macroplasma) coating method renders plenty of disadvantages in terms of phase impurity, lesser porosity that hinders osseointegration, and residual stresses, which ultimately lead to inadequate mechanical properties and delamination of the coating. More often than not, such MAPS-HAp coatings may possess HAp or other Ca-P phases, e.g., tricalcium phosphate (TCP), tetracalcium phosphate (TTCP), as well as an amorphous phase that leads to poor crystallinity. As a result, properly controlled bioresorption, which could have been achieved by the presence of the pure HAp phase, can be adversely affected. That is, the process could become either unacceptably fast or devastatingly slow, in which in both cases the ultimate result becomes the poor efficacy of such MAPS coatings.

This global scenario leads us to think in a different manner to arrive at a better solution to the generic problems associated with such coatings. Therefore, we have utilized a low-power plasma spraying, i.e., the microplasma spraying (MIPS) method. The MIPS method has been developed very recently, e.g., only in the mid-1990s onward. In the current research, this MIPS technique has been utilized to coat HAp on SS316L substrates to minimize the aforesaid problems associated with conventional, commercial macroplasma sprayed hydroxyapatite coatings. Looking at the concept of the affordable healthcare problem for the large, ailing aged male and female populations usually subject to the need of HAp-coated biomedical prosthesis implantation, especially in the third world and developing countries, surgical-grade SS316L has been given more importance as a substrate material because it is more cost-effective than Ti-6Al-4V and CoCrMo alloys. In this book, systematic studies are discussed from the processing of the MIPS-HAp coating to its physicochemical and mechanical properties at both the macro- and microstructural length scales. In addition, the physical, structural, micromechanical, and micro- and nanotribological properties of the MIPS coatings were evaluated as well as analyzed following exposure to simulated body fluid (SBF) solutions for about 2 weeks of immersion time. Based on the successful performance of the developed MIPS coatings under an accelerated condition of exposure to SBF, it was actually put into an in vivo application in an animal model, e.g., the MIPS-HAp coating was applied on a metallic intramedullary pin used for bone defect repair in a New Zealand white rabbit. How the MIPS-HAp-coated intramedullary pin performed much better than the uncoated metallic pin is discussed. Apart from these aspects of specific development, an in-depth discussion is provided on the comparative pros and cons of the other processing techniques used to develop bioactive HAp coatings in general. Emphasis has been generally dedicated to dig out the correlation between the processing parameters and the resultant microstructure on the one hand, and the mechanical and tribological properties of the coatings on the other. Further, the general historical evolution and the scientific advantages of the microplasma spraying technique as a unique advancement or biomedical, societal need-driven modification of conventional high-power plasma spraying have been amply and systematically documented. To the best of the authors' knowledge, this book in the area of microplasma sprayed HAp coating is the first such attempt, and we hope that it marks the beginning of a journey into this area of research.

What Is in It for Us?

This book is divided into 12 chapters intertwined with one another. Chapter 1 deals with the introduction of the problem, such as what a biomaterial really is, what the frequently used term *biocompatibility* stands for in general, and in particular why it is so important that the biomaterials be biocompatible. This discussion unfolds into the types of biomaterials and their characteristic features, followed by a more dedicated discussion on bioceramics, especially HAps and specifically HAp coatings. In the process of such discussions, two most important topics are touched upon. The first is a brief take on natural biomaterials, e.g., bone and teeth, as they almost generically provide the largest application areas for HAp and HAp coatings. The second one that has been glanced upon is why and how surface engineering of implants has evolved as an important area of research and development in the journey to provide better biomaterials for society today, what roles could possibly be played by such synthetic biomaterials and coatings in such developments, and to attain such improvements, what challenges are to be faced by the community.

Chapter 2 deals with coating techniques, with a special emphasis on plasma spraying techniques. In particular, the basics of coating formation by the plasma spraying technique are discussed, with an emphasis on uniqueness and applications of microplasma spraying as a manufacturing technique in relation to the developmental needs of plasma-sprayed HAp coatings. This scenario is placed in contrast with the other, more conventional coating techniques, such as high-velocity oxy fuel, pulsed laser deposition, cold spraying, liquid precursor plasma spraying, electrophoretic deposition, electrohydrodynamic atomization, physical vapor deposition techniques such as vacuum sputtering, radio frequency sputtering, reactive sputtering, etc., sol-gel and dip coating, biomimetic coating process, etc. Finally, the conventional macroplasma spraying (MAPS) process and the newly developed, relatively greener, and more utilizable and cost-effective microplasma spraying (MIPS) process are amply compared to highlight their relative merits and demerits.

Chapter 3 essentially focuses on the materials science aspect of HAp coating and its applications. Starting with the background of the problem faced in fixation of biomedical implants and the basic scientific issues involved in the journey toward achieving a fruitful solution to this problem, the HAp coatings developed by different methods are compared in order to provide a state-of-the-art scenario. The issues related to high- and low-temperature processing of coatings in general are discussed, with a special emphasis on elucidating the roles of those processing parameters that are of paramount importance in the development of any coating by the MIPS technique. In addition, the pros and cons of MAPS- and MIPS-HAp coatings are adjudged with a view to justify the choice of the MIPS technique as useful for HAp coating development. Thereafter, the roles of various processing parameters, e.g., plasma spraying atmosphere, spraying current and stand-off distance, gun traverse speed, specimen holder arrangements, on plasma spraying of HAp coatings are illustrated, emphasizing the efficacy of the MIPS as a suitable method for HAp coating development. Global efforts to combat brittleness of HAp coatings through nano-HAp and HAp composite developments are highlighted. The chapter concludes with a major focus on current scenarios in research and development of HAp coatings.

Chapter 4 stresses structural and chemical properties of HAp coatings. It basically deals with the techniques involved in characterizing the microstructural parameters, e.g., thickness, porosity, crystallinity, stoichiometry, phase purity, surface chemical structure, splat size, micro- and macropore sizes and their distribution, micro- and macrocrack sizes and their distribution. Next, an account is provided of the variations in the aforesaid microstructural parameters when measured as sprayed and polished HAp coatings, with particular emphasis on the formation and growth aspects of the microstructure itself, along with its constituent parameters. A rational explanation of anisotropy of Young's modulus values measured in directions parallel and perpendicular to the directions of spraying of the HAp coating is provided next. Subsequently, comprehensive modeling efforts are presented to justify the dependencies of Young's modulus on various pore shapes present in a given HAp coating. Finally, the predictions of the modeling efforts are validated by experimental data measured by the nanoindentation technique applied to both cross section and plan section of a given HAp coating.

The focus of Chapter 5 is kept on the in vitro properties of hydroxyapatite coatings. Starting from a comprehensive survey of existing literature data of relevance, this chapter provides a detailed comparison of the efficacies under SBF immersion for both MAPS- and MIPS-HAp coatings developed on biomedical implant-grade SS316L substrates.

Explicit coverage on the macromechanical and micro/nanomechanical properties of HAp coatings is provided in Chapters 6 and 7, respectively. Chapter 6, in particular, focuses on issues related to factors that govern the bonding strength of HAp coatings, connected parameters, relevant measurement techniques, standards of measurements, influences of microstructure, and residual stress on bonding strength of MAPS- and MIPS-HAp coatings, as well as on those of HAp coatings developed by techniques other than MAPS and MIPS, and so on. Finally, from the perspective of in-service reliability of HAp coatings, the importance of evaluating the shear strength, push-out strength, flexural strength, and fatigue behavior is adequately discussed.

It needs to be emphasized here that the coating should have structural integrity not only at the macrostructural length scale, but also at the microstructural length scale. This is what gives Chapter 7 its genesis and scope to flourish. It starts with the basics of the nanoindentation technique, which plays a crucial role in evaluations of nanohardness, Young's modulus, and for the very first time, the fracture toughness of MIPS-HAp coatings in the presence or absence of an SBF immersion history. After presenting the limitations in literature scenario on evaluations of nanohardness and Young's modulus of the HAp coatings, the applications of the unique nanoindentation technique for evaluations of nanohardness and Young's modulus of the HAp coatings are focused in detail here. The issue of statistical reliability and how to tackle the characteristic scatter in the experimentally measured nanohardness and Young's modulus of the intrinsically heterogeneous HAp coatings is discussed in utmost detail, followed by a rational explanation about the achievement of great toughness in these coatings.

It needs to be emphasized further here that it is not only important for a given HAp coating to have a reasonably good, acceptable combination of quasi-static mechanical properties at both macrostructural and micro-structural length scales, but also important to have equally good, accept-able dynamic contact resistance, e.g., micro- and nanotribological properties, simply because the biomedical prosthesis will be subject to both quasi-static and dynamic contact situations when in service. That is why the subject of Chapter 8 is the tribological properties of the hydroxyapatite coatings. The typical survey of existing literature does indeed focus on serious limitations in both the database and knowledge base on tribological behavior of HAp coatings in general and MIPS-HAp coatings in particular. Thus, the micro-as well as nanotribological behavior of both as-sprayed and polished MIPS-HAp coatings before and after the SBF immersion treatments are adequately discussed here to adjudge its efficacy further for, say, in vivo trials in animal models. It is also recognized that a biomedical prosthesis such as a HAp-coated implant will be exposed to repeated loading and unloading during in-service usage. Such a situation can lead to fatigue, and if the coating is not fatigue resistant, then to its premature failure. This is the situation that is the least desirable. That is why this chapter concludes with a glimpse on a litera-ture scenario about the fatigue behavior of various HAp coatings.

Let us assume for argument's sake that a given HAp coating has good mechanical properties at both macrostructural and microstructural length scales, in addition to having acceptable tribological behavior. Will this guar-antee that its in-service performance will be failure-free? Unfortunately, the response to such a query will not be affirmative because in spite of having all the aforesaid attributes, the coating can still fail due to the most undesirable but absolutely unavoidable presence of residual stress. From the perspective of practical applications of HAp coatings in general and MIPS-HAp coatings in particular, therefore, the issue of residual stress is of paramount impor-tance, as dealt with in all possible detail in Chapter 9. Starting from a compre-hensive survey of the relevant literature scenario to provide a state-of-the-art account of residual stress in HAp coatings, it focuses on the pros and cons of the various conventional methods for residual stress estimations before pro-viding a picture of how the related processing parameters of various coat-ing deposition techniques in general and plasma spraying in particular can affect the residual stress scenarios in both global and local microstructural scenarios of a given coating. Finally, in conclusion, the efficacy of a novel nanoindentation technique in successfully assessing the nature and magnitude of the residual stress in a given MIPS-HAp coating is established.

The first-order acid test of a given bioceramic coating lies in its success in animal models during in vitro studies. Thus, Chapter 10 focuses precisely on in vivo performance efficacy examination of MIPS-HAp coatings in both rat and goat models. The subject matter of this chapter therefore mainly comprises the results of a pioneering study on the comparative healing performance efficacies of uncoated and MIPS-pure HAp-coated intramedullary SS316L pins for bone defect fixation in New Zealand white rabbits. It has indeed been interesting to note how biochemical, histological, radiographic, and fluorochrome labeling studies could be used to establish comparatively much more "bone bonding" efficacy and osseointegration in HAp-coated intramedullary SS316L pins, which has led to complete bone defect healing that was not possible to attain with the uncoated intramedullary SS316L pins.

Finally, Chapter 11 introduces the futuristic scopes and colors on the horizons for more challenging and useful developments that are on the verge of emergence in the R&D scenario of MIPS-HAp coatings. It also links up these concepts to further establish improving the quality of human life in the future; such developments are of utmost utility to the biomedical research community in general and the bioactive ceramic coating research community in particular.

As presented in Chapter 12, the book ends with a summary of what has been achieved through these humble efforts.

Arjun Dey
Anoop Kumar Mukhopadhyay

Acknowledgments

At the onset, we express our sincere thanks to the directors of CSIR–Central Glass and Ceramic Research Institute (CSIR CGCRI), Kolkata, and ISRO Satellite Centre, Bangalore, for their constant encouragement. We are grateful to Mr. Srikanta Dalui, Mr. Manas Raychoudury, Mr. Haradhan Das, Mr. Ram Narayan Kumar, Mr. Kartik Banerjee, and Mr. Biplab Naskar of the Mechanical Property Evaluation Section, CSIR-CGCRI, for help received during this work. We are especially indebted to the late Dr. D. Basu, Dr. M. K. Sinha, Mr. B. Kundu, and Mr. Anath Karmakar from the Bio-Ceramics and Coatings Division of CSIR–CGCRI.

What Arjun Says

It is now my turn to thank the people who were involved behind the curtain for the completion of this book. First, I am really grateful to my beloved parents, Mr. Dilip Kumar Dey and Ms. Minati Dey, and to my dear wife, Omprita Chatterjee, who took on all the responsibilities for family matters and gave me the encouragement to publish this book. I am also most thankful to my dear boudi (sister-in-law) Ms. Manjulekha Mukherjee (spouse of Dr. Mukhopadhyay) for providing continuous mental support and caring for almost every need during the publication of this book.

Very special thanks are due to Dr. A. K. Sharma, Group Director, Thermal System Group, ISAC-ISRO; Prof. Bikramjit Basu, IISc Bangalore; Prof. Tapas Laha, IIT Kharagpur; Prof. Srinivasa Rao Bakshi, IIT Madras; Prof. Vikram Jayram, IISc; Prof. Rabibrata Mukherjee, IIT Kharagpur; Dr. Harish Barshilia, CSIR-NAL; Prof. Niloy K. Mukhopadhyay, IIT-BHU; Prof. Kuruvilla Joseph and Prof. K. Prabhakaran, IIST, Trivundrum; Prof. Subroto Mukherjee, IPR, Gandhinagar; Dr. Arvind Sinha, CSIR-NML; Prof. Ajoy Kumar Ray, Director, IIEST, Shibpur; and Prof. Gerhard Wilde, Director, Institute of Materials Physics, University of Muenster. Moreover, I have always received enormous support from Prof. Nil Ratan Banyopadhyay and Prof. Subhabrata Datta from IIEST, Shibpur, without whose support my academic and research career would not, perhaps, have grown to where it is today. Before I finish, I am really fortunate to have met Dr. Anoop, who is not only my master's and PhD guide and coauthor of this book, but also the pathfinder of my research life.

Arjun Dey

What Anoop Says

I thank my late, beloved parents, Mr. Girija Bhusan Mukherjee and Mrs. Kamala Mukherjee, and my late in-laws, Mr. Beni Madhab Bhattacharya and Mrs. Bijaya Bhattacharya, for making me what I am today. I wish all of them were alive today! The huge support received from Mr. and Mrs. S. K. Mukherjee of Hindustan Aeronautics Ltd., Bangalore; Prof. S. Bhattacharya of Institute of Management Technology, Ghaziabad; Dr. Mrs. A. Bhattacharya of SDGI, Ghaziabad; and Mr. and Mrs. A. Bhattacharya of Silvasa, Mumbai, are also gratefully acknowledged. It would be an oversight not to recognize the tremendous contributions of all four of my sisters-in-law, four late brothers and one late sister, and two late uncles and aunts to my upbringing and their huge encouragement of my academic efforts.

I am also more than grateful for my entire life to my dear, beloved wife, Mrs. Manjulekha Mukherjee, and to my two sweet daughters, Miss Roopkatha Mukhopadhyay, aged 17, and Miss Sanjhbati Mukhopadhyay, aged 12; without their tremendous patience, love, care, and mental support, this book would not have seen the light of the day. It is a joy for me to recognize the kind encouragement received from my nephew and his wife, Mr. N. Mukherjee and Mrs. M. Mukherjee, and my niece, Miss L. Mukherjee.

I also appreciate very much the kind contributions made by Profs. Y.-W. Mai, M. V. Swain, and Mark Hoffman of Australia; Dr. Brian Lawn of NIST, United States; Prof. S. Priyadarshy of the United States; Dr. R. W. Steinbrech of Germany; Mr. I. Dean of South Africa; Prof. M. K. Sanyal, ex-Director, SINP; Dr. D. Chakraborty, my own PhD guide; Dr. S. Kumar, Dr. B. K. Sarkar, Dr. C. Ganguly, Dr. H. S. Maity, Dr. K. K. Phani, Dr. D. K. Bhattacharya, and late Drs. A. P. Chatterjee, D. Basu, and Mr. D. Chakraborty, all of CSIR-CGCRI; Dr. S. Tarafder of CSIR–National Metallurgical Laboratory, Jamshedpur; Prof. B. Basu of IISc, Bangalore; Prof. N. R. Bandyopadhyay of IIEST, Shibur; Prof. A. N. Basu of Jadavpur University; and of course, Prof. I. Manna of IIT Kanpur to my development as a researcher. I especially thank Prof. R. Mukherjee of IIT Kharagpur and the late Mr. P. Basu Thakur of Icon Analytical Pvt. Ltd., India, without whose very active support the Nanoindentation Laboratory would not have been created at CSIR–CGCRI. I also thank my colleagues Mr. D. Sarkar, Mr. I. Biswas, Drs. D. Bandyopadhyay and G. Banerjee, Miss S. Datta, Mr. S. Dey, Mrs. S. Nandan, and Mr. S. Biswas of the Planning and Project Management Division, and all my other colleagues from the Publication Section and the Non-Oxide Ceramic and Composite Division of CSIR–CGCRI, for their help and cooperation during the course of writing this book.

I am also very grateful to my dear brothers, Mr. S. Acharya and Mr. D. Moitra of CSIR-CGCRI, and to Mr. R. Das, as well as Dr. Anup Khetan of RTIICS, without whom I would not have been able to write even the very first page of this book. I also thank Dr. S. K. Bhadra, Chief Scientist, and

Mr. K. Dasgupta, Director of CSIR-CGCRI, for their kind encouragement, support, and advice at all stages. Very special thanks are due to my colleagues Dr. R. N. Basu, Dr. D. Kundu, Dr. G. De, Dr. H. S. Tripathy, Dr. S. Dasgupta, Dr. V. K. Balla, Dr. S. K. Ghosh, and Mr. A. K. Chakraborty of CSIR-CGCRI. I also thank the Almighty to have kindly provided me a student such as Dr. Arjun Dey, the coauthor of this book. He and his wife, Omprita, are part of my life and definitely much more than mere students or junior colleagues to me. I feel very proud to have the opportunity to be his thesis supervisor, just by chance, as they say, by the decree of God. I acknowledge all the near and distant relatives, teachers, friends, colleagues, former and present students, and well-wishers for their support and kind encouragement received at every bend of my life, and particularly during the execution of this book writing project. Last but not the least, I express my gratitude to my dear friend, the poet, Joy Goswami, who taught me how to live through the hours of grief and pain, through the hours of tortures and torments, just to see through the apparently endless night that ends in a new whisper of love, a new breath of life, and the sunrise of a new day.

Anoop Kumar Mukhopadhyay

About the Authors

Dr. Arjun Dey is presently working as a scientist at the Thermal Systems Group of ISRO Satellite Centre (ISAC), Indian Space Research Organisation, Department of Space, Bangalore. Dr. Dey earned a bachelor's degree in mechanical engineering in 2003 from Biju Patnaik University, Orissa, followed by a master's degree in materials engineering from the Indian Institute of Engineering Science and Technology, Shibpur, Howrah (formerly Bengal Engineering and Science University), in 2007. While working at CSIR–Central Glass and Ceramic Research Institute (CGCRI), Kolkata, he earned his doctoral degree in materials science and engineering in 2011 from the Indian Institute of Engineering Science and Technology, Shibpur, Howrah. His PhD thesis was titled *"Physico-Chemical and Mechanical Characterization of Bioactive Ceramic Coating."*

He has been given many prestigious awards, such as the Dr. R. L. Thakur Memorial Award from the Indian Ceramic Society in 2012, the DST-Fast Track Young Scientist Scheme Project Grant Award from the Department of Science and Technology, India, in 2011, the Young Engineers Award and Metallurgical and Materials Engineering Division Prize from the Institution of Engineers (India) in 2010–2011, and the Young Scientist Award from the Materials Research Society of India, Kolkata Chapter, in 2009, for significant contributions in the fields of materials science and engineering. Arjun also obtained the prestigious CSIR–Senior Research Fellowship Award from the Council of Scientific and Industrial Research (CSIR) to accomplish his PhD. The research work of Dr. Dey culminated in more than 150 publications to his credit. He serves as a reviewer for many national and international journals. Dr. Dey has chaired a technical session on *"Surface Engineering, Thin Films and Coatings"* in the 3rd Asian Symposium on Materials and Processing (ASMP-2012) held at IIT Madras, Chennai, August 30–31, 2012. Recently, he has coauthored the book *Nanoindentation of Brittle Solids*, published by the Taylor and Francis Group/CRC Press in June 2014.

Dr. Anoop Kumar Mukhopadhyay is a chief scientist and head of the Advanced Mechanical and Materials Characterization Division of CSIR–Central Glass and Ceramic Research Institute (CSIR–CGCRI), Kolkata, India. He earned his bachelor's degree with honors in physics from Kalyani University, Kalyani, in 1978, followed by a master's degree in physics from Jadavpur University, Kolkata, in 1982. In 1982, he initiated in India research on microstructure-mechanical property correlaton of non-oxide ceramics. These ceramics were meant for high temperature application. He joined CSIR-Central Glass and Ceramic Research Institute as a staff scientist in 1986. Working on the critical parameters that control the high-temperature fracture toughness of silicon nitride and its composites, he earned his PhD degree in science in 1988 from the Jadavpur University, Kolkata.

He was awarded during 1990–1992 the prestigious Australian Commonwealth Post Graduate Research Fellowship to do post-doctoral work at the University of Sydney, Australia. During this post-doctoral work he worked with world renowned professors Yiu-Wing Mai and Michael V. Swain at the same university. His post doctoral work involved development of wear and fatigue resistant ceramics. During this period he made pioneering contribution on the role of grain size in the wear mechanism of alumina ceramics. At CSIR–CGCRI, Kolkata, Dr. Mukhopadhyay established an enthusiastic research group on the evaluation and analysis of mechanical and nanomechanical properties of glass, ceramics, bioceramic coatings and biomaterials, thin films, and natural biomaterials.

Dr. Mukhopadhyay has authored more than 200 publications including well over 100 papers in international, peer-reviewed SCI journals. He has authored seven patents, with four of them already granted, two book chapters already published, and two books (one published in May/June 2014, one in progress). He has supervised seven doctoral students. He contributed three chapters in the *Handbook of Ceramics*, edited by Dr. S. Kumar, internationally famous glass technologist and former director of CSIR–CGCRI, Kolkata, India, and published by Kumar and Associates, Kolkata. He serves on the editorial board of *Soft Nanoscience Letters*. In 2008, he won the Best Poster Paper Award at the 53rd DAE Solid State Physics Symposium. He also won, in 2000, the Sir C. V. Raman Award of the Acoustical Society of India. In the same year, he won the Best Poster Paper Award of the Materials Research Society of India. He was also awarded in 2000 a Visiting Scientist Fellowship to work on the fracture and nanoindentation behavior of ceramic thermal barrier coatings with world-renowned scientist Dr. R. W. Steinbrech at the Forschungszentrum, Juelich, Germany. He was awarded in 1997 the Outstanding Young Person Award for Science and Innovation by the Outstanding Young Achievers Association, Kolkata, and won the Lions Club of India award in 1996. His work was recognized in 1995 through the Best Poster Paper Award of the Materials Research Society of India. Recently, in 2010, his paper won the Best Research Paper Award at the Diamond Jubilee Celebration Ceremony of CSIR–CGCRI, Kolkata. Very recently in 2013 he was awarded the prestigious Materials Research Society of India Medal by the Materials Research Society of India.

Recently, he coauthored, with Dr. Arjun Dey, his former PhD student and a present colleague from ISAC, ISRO, Bangalore, the book *Nanoindentation of Brittle Solids*, published by CRC Press/Taylor & Francis Group.

His current research interests cover a truly diverse span, e.g., physics of nanoscale deformation for brittle solids, very high strain rate shock physics of ceramics, tribology of ceramics, nanotribology of ceramic coatings and thin films, microstructure mechanical and functional property correlation, as well as ultrasonic characterization and fatigue of (1) structural and bioceramics, bioceramic coatings, and biomaterials; (2) multilayer composites; and (3) thick/thin hard ceramic coatings. He also has a very active interest in microwave processing of ceramics, ceramic composites, and ceramic–metal or ceramic–ceramic joining.

Common Abbreviations

APS	Atmospheric plasma spraying
BCP	Biphasic calcium phosphate
CaP	Calcium phosphate
C/C	Carbonated fiber reinforced carbon matrix composites
CHAp	Carbonate hydroxyapatite
CNT	Carbon nanotube
COF	Coefficient of friction
CTE	Coefficient of thermal expansion
DEJ	Dentin and enamel junction
DLC	Diamond-like carbon
E	Young's modulus
EDX	Energy-dispersive X-ray spectroscopy
EHDA	Electrohydrodynamic atomisation
EPD	Electrophoretic deposition
E-SEM	Environmental scanning electron microscopy
FCS	Fetal calf serum
FESEM	Field emission scanning electron microscopy
FHAp	Fluoridated HAp
FTIR	Fourier transform infrared spectroscopy
H	Hardness/nanohardness
HAp	Hydroxyapatite
HBP	Human blood plasma
HBSS	Hanks' balanced salt solution
HVOF	High-velocity oxy fuel
ICP-AES	Inductively coupled plasma atomic emission spectroscopy
IFM	Indentation fracture method
IP	Intellectual properties
ISE	Indentation size effect
LPPS	Liquid precursor plasma spraying
MAPS	Macroplasma spraying
MIPS	Microplasma spraying
MWCNT	Multiwall carbon nanotube
nHAp	Nanohydroxyapatite
OCP	Octacalcium phosphate
p	Open porosity
$P\text{-}h$	Load vs. depth
PLD	Pulsed laser deposition
PMMA	Polymethylmethacrylate
PVD	Physical vapor deposition
rf	Radio frequency

S	Shear strength
SBF	Simulated body fluid
SEM	Scanning electron microscopy
SG	Sol-gel
SPM	Scanning probe microscopy
SWCNT	Single-wall carbon nanotube
TCP	Tricalcium phosphate
TEM	Transmission electron microscopy
TTCP	Tetracalcium phosphate
VE	Vacuum evaporation
VPS	Vacuum plasma spraying
XRD	X-ray diffraction
YSZ	Yttria-stabilized zirconia
ZTA	Zirconia-toughened alumina

1

Introduction

In the present chapter, the basics of biomaterials, the types of biomaterials in general and the bioactive materials in particular, will be discussed. The different categories of bioceramics will be described. However, particular emphasis will be on the usage of bioactive hydroxyapatite and its coating. The naturally occurring hydroxyapatite is found in bone and teeth. Therefore, these natural nanocomposites shall be briefly discussed, putting in perspective their structural complexity on the one hand, and their marvelous functionality on the other. However, before going to such detail, we need to know what we really mean by the term *biomaterial*, as discussed below.

1.1 Introduction of Biomaterials

1.1.1 What Is a Biomaterial?

What is a biomaterial? Classically speaking [1], any substance (other than drugs), synthetic or natural in origin, which can be used for any period of time (long or short term), as a whole or as a part of a system, that treats, augments, or replaces any tissue, organ, or function of the body is typically defined as a biomaterial. The same functionality can also be provided by a combination of different materials, which in such case form the generic group biomaterials. Such materials are designed to replace a part or a function of the human body in a safe, reliable, economic, and physiologically as well as aesthetically acceptable manner. Any synthetic material incorporated into a human organism has to abide by certain properties that will ensure that there are no negative interactions with living tissue [1]. On the other hand, a biological material is a natural material such as skin, artery, liver, etc., produced by a biological system. The performance of a biomaterial is bound to be compared to its possible natural counterpart, i.e., the biological material, and that is where the challenge lies in the development of such materials.

1.1.2 What Is Biocompatibility?

We often hear the term *biocompatibility*. What does it really mean? Well, even before we understand what biocompatibility is, we need to understand the concept of it. Basically, if we look at the other part of this term, that apart from *bio*, we have *compatibility*. It's like, if we may imagine, a new student coming into an ongoing class. Does he quarrel with everybody? Is he friendly with everybody? Does he mix up well with the other students? Does he remain as a disturbing outsider? Or, is he an indifferent outsider? The answers to all these questions and the interactive steps involved in arriving at such answers really define to what extent the new student is compatible with the ongoing class batch. Just like this, biocompatibility refers to the ability of a material to perform with an appropriate host response (i.e., the response of the host organism to the implanted material or device) in a specific application. That is why biocompatibility is neither a single event nor a single phenomenon. Rather, it is meant to be a collection of processes involving different but interdependent interaction mechanisms that may be undergone between the living tissues and a given material [2].

1.2 Types of Biomaterials

The classification of biomaterials appropriately follows the classical conventional categories of structural materials such as metals (e.g., SS316L, Ti, Ti-6Al-4V, NiTi, CoCrMo alloy, etc.), ceramics (alumina, zirconia-toughened alumina (ZTA), hydroxyapatite (HAp), tricalcium phosphate (TCP), etc.), and bioglasses polymers (polyethylene, polymethylmethylacrylate (PMMA), etc.), and composites [3]. In increasing order of biocompatibility, the interaction of biomaterials with living tissue can be defined as follows [4, 5]:

1. Incompatible
2. Biocompatible
3. Nearly bioinert
4. Bioactive

1.2.1 Incompatible and Biocompatible Materials

Incompatible materials release to the body substances in toxic concentrations or trigger the formation of antigens that may cause immune reactions ranging from simple allergies to inflammation to septic rejection with the associated severe health consequences. The biocompatible materials also release substances, but in nontoxic concentrations that may lead to only benign tissue reactions, such as formation of a fibrous connective tissue capsule or weak immune reactions that cause formation of giant cells (i.e., phagocytes).

These materials are often called biotolerant and include austenitic stainless steels (SS316L) or bone cement consisting of PMMA.

1.2.2 Nearly Bioinert Materials

However, nearly bioinert materials do not release any toxic constituents and do not show positive interaction with living tissue. As a response of the body to these materials, usually a nonadherent capsule of connective tissue is formed around the bioinert material. In the case of bone remodeling, it manifests itself by a shape-mediated contact osteogenesis [6]. Compressive forces are transmitted ("bony on-growth") only through the bone-material interface. Typical bioinert materials are titanium and its alloys, ceramics such as alumina, zirconia, and ZTA, and some polymers as well as carbon. They do not possess any chemical bonding with the host tissue.

1.2.3 Bioactive Materials

Finally, bioactive materials show a positive interaction with living tissue. Such interaction typically also includes the differentiation of immature cells toward bone cells. In this case of course, there is a chemical bonding to the bone along the interface. This is thought to be triggered by the adsorption of bone growth-mediating proteins at the biomaterial's surface. Hence, a biochemically mediated strong bonding osteogenesis is initiated [6]. As a result, mostly the compressive and, to a minor extent, the tensile as well as shear forces can also be transmitted through such an interface (bony in-growth). Typical examples of the bioactive materials are calcium phosphates and bioglasses. It is believed that bioactivity of calcium phosphates is associated with the formation of carbonated hydroxyapatite (CHAp), similar to bone-like apatite [7].

1.3 Categories of Bioceramics

There is another viewpoint advocated in literature. Thus, three broad categories of bioceramics, e.g., nearly inert, surface reactive, and resorbable, have also been defined (Figure 1.1) [6, 8]. Such categories are based on the chemical reactivity of the bioceramics in an in vivo environment. On a relative scale, nearly inert bioceramics are, e.g., alumina, zirconia, ZTA, etc. They tend to exhibit inherently low levels of reactivity, which peaks at about 10,000 days. Surface-reactive bioceramics, such as HAp, Hench's bioglass [9], etc., have a substantially higher level of reactivity, which peaks at about ~100 days.

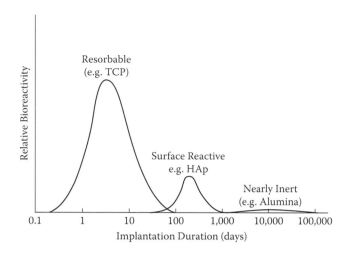

FIGURE 1.1
A schematic of chemical reactivity in in vivo environments of different bioceramics. Based on
that, bioceramics are categorized in three subgroups. (Adopted and modified/redrawn from
Hench, *American Ceramic Society Bulletin*, 72: 93–98, 1993; and Shackelford, *Bioceramics*, vol. 1.,
Amsterdam, The Netherlands: Gordon and Breach Science Publishers, 2005.)

Resorbable bioceramics, such as TCP, have even higher levels of reactivity
that peak at about ~10 days. This broad spectrum of chemical behavior has
led to a corresponding range of engineering design philosophies.

1.4 What Is Hydroxyapatite?

Hydroxyapatite (HAp) is chemically similar to the primary mineral content
of bone and teeth [6, 10]. Chemically, it is represented as $Ca_{10}(PO_4)_6(OH)_2$. It
is characterized by a Ca/P ratio of 1.667. However, it is difficult to obtain the
exact stoichiometric ratio (Ca/P ratio ~ 1.667) in HAp because of the different
Ca/P ratios that can be stabilized, depending on the synthesis method and
conditions employed. Nevertheless, HAp has the desirable physicochemical attributes of stability, inertness, and biocompatibility. Therefore, HAp
is classified as bioactive. This means that it will support bone ingrowth and
osseointegration when used in orthopedic, dental, and maxillofacial applications. Various HAp phases can be formed, which can be categorized into
calcium-deficient hydroxyapatite, oxy-hydroxyapatite, and CHAp, depending on the type of environment employed in the synthesis steps. The chemical nature of HAp lends itself to substitution. As a consequence, it is not
uncommon for nonstoichiometric hydroxyapatites to exist. The most common substitutions involve carbonate, fluoride, and chloride substitutions

for hydroxyl groups. Such processes form the deficient hydroxyapatites. However, biological apatites differ chemically from ideal HAp. Then the almost obvious question is: What is the difference? Well, the biological apatites often include cations such as Mg^{2+}, Na^+, and K^+ and anions such as Cl^- and F^-. In principle, such inclusions can be introduced into the HAp lattice by substitution of one or more Ca^{2+}. Nevertheless, the major constituent in biological apatite is carbonate. In bone mineral the carbonate is present by ~5–8 wt% [11, 12]. There is another group of calcium phosphate (CaP) ceramics called biphasic calcium phosphate (BCP) bioceramics. These belong to a group of bone substitute biomaterials. Such biomaterials comprise of an intimate mixture of varying HAp/β-TCP ratios. The general merits and demerits of HAp are [1, 5, 11]:

1. Without breaking down or dissolving, it integrates into the bone structures and supports bone in-growth.

2. It possesses a thermally unstable compound, decomposing from about 800–1250°C, depending on its stoichiometric ratio.

3. It does not have the requisite mechanical strength to guarantee its success in long-term load-bearing applications.

1.5 What Is Hydroxyapatite Coating?

Therefore, instead of being used as a bulk compact, HAp is generally used as a coating on bioinert metallic implants. The bond formed between a metallic alloy implant and the bone tissue is mediated by a so-called contact osteogenesis process, as explained earlier. Bone tissue grows one-directionally toward the interface and bony on-growth occurs. Thus, the bony on-growth is able to only transmit the compressive loads. However, in real-life conditions, the actual loads that the interface is subjected to during movement of the patient also contain strong tensile and shear components that will have to be taken care of by the implant. Therefore, in the clinical practice, in many cases a bioactive HAp layer is provided. Such a process allows bonding osteogenesis through bony ingrowth. As a result, such a strong interface displays the ability to transmit these tensile and shear forces [5, 11, 13].

In such a bonding osteogenesis process, two ossification fronts develop. One grows from the bone toward the implant. The other grows from the implant toward the bone [5, 11]. Evidence is mounting that a 150 μm long-term stable bioactive HAp coating will elicit a specific biological response at the interface of the implant material by control of its surface chemistry through adsorption of noncollageneous proteins such as osteocalcin, osteonectin, silylated glycoproteins, and proteoglycanes [11]. This process results in the eventual establishment of a strong and lasting osseoconductive bond between living tissues and the biomaterial. The bioactive HAp coatings possess some unique properties [11]; for example:

1. It prevents the formation of a fibrous capsule of connective tissue surrounding the metallic implant.
2. It achieves fast bone apposition rates through preferential adsorption of proteins.
3. It generates a bonding osteogenesis that provides a continuous and strong interface between metallic implant and tissue that is able to transmit not only compressive but also tensile and shear loads.
4. It develops accelerated healing compared to those of the implants without a bioactive coating.
5. It achieves a decrease in the release of metal ions to the surrounding tissue, a process that can minimize the perceived risk of a cytotoxic response.

The data [14] on the variations of interfacial bond strength of both the HAp-coated and uncoated metallic implants as a function of implantation time are shown in Figure 1.2. Interfacial bond strength of the HAp-coated implants was always higher than those of the bare implants [14].

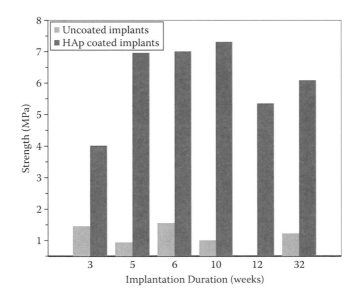

FIGURE 1.2
Variations of interfacial bond strength of both the HAp-coated and uncoated metallic implants as a function of implantation time. (Redrawn from Hench, *Journal of the American Ceramic Society*, 74: 1487–1510, 1991. From American Ceramic Society and Wiley.)

1.6 Introduction of Bone: A Natural Biomaterial

Bone is a natural ceramic-polymer hybrid composite. It consists mainly of collagen (20%), calcium phosphate, e.g., HAp (69%), and water (9%). Other organic substances, e.g., proteins, polysaccharides, and lipids, are present in small amounts [15]. Collagen is located in bone tissue. It has the form of fibrils 100–2000 nm in diameter. Calcium phosphate in the form of crystallized HAp ensures bone rigidity. The HAp crystals have the shape of needles. These needles are ~40–60 nm long, ~20 nm wide, and ~1.5–5 nm thick [15–20]. The mineral component of bone is similar to HAp. However, it contains in addition fluoride, magnesium, sodium, and other ions as impurities. The weight percents of the elements are as follows: [18] calcium, 34.8%; phosphorus, 15.2%; sodium, 0.9%; magnesium, 0.72%; potassium, 0.03%; carbonates, 7.4%; fluorine, 0.03%; chlorine, 0.13%; pyrophosphates, 0.07%; and other elements, 0.04%.

Bone is rather nonuniform (i.e., anisotropic) in microstructure and mechanical properties. The mechanical properties of bone show strong sensitivity to porosity (5–95%), the degree of mineralization, and the orientation of the collagen fibers [19]. Cortical bone is a nanostructured composite. It is made up of HAp-based matrix and collagen fibers. The matrix has a layered microstructure. This matrix provides the basis for oriented cylindrical formations on a macroscopic scale [20]. The length scale is very important in describing the hierarchical architecture of bone and understanding relationship between structures at various levels of hierarchy [17].

As illustrated in Figure 1.3, these levels and structures are:

1. The macrostructure comprising cancellous (porous) and cortical (hard) bone
2. The microstructure (from 10 to 500 μm) comprising Haversian systems, osteons, and a single trabecula
3. The submicrostructure (1–10 μm) comprising lamellae
4. The nanostructure (from a few hundred nanometers to 1 μm) comprising fibrillar collagen and embedded mineral
5. The subnanostructure (below a few hundred nanometers) comprising the molecular structure of constituent elements, such as mineral, collagen, and noncollagenous organic proteins

The scanning electron photomicrograph of lamellar bone structure is shown in Figure 1.4. This hierarchical organized structure has an irregular, yet optimized arrangement and orientation of the components, making the material of bone heterogeneous, anisotropic, and tough. This complex hybrid structure is responsible for the high strength and fracture toughness

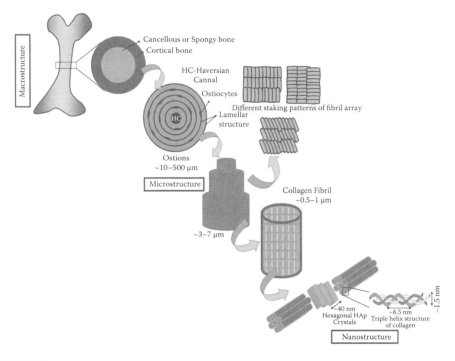

FIGURE 1.3
Schematic of the hierarchical structure of bone.

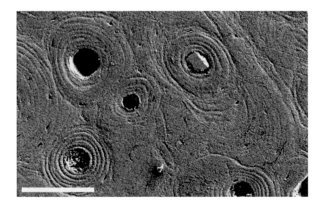

FIGURE 1.4
Scanning electron photomicrograph of lamellar structure of human cortical bone (scale bar indicates 100 μm).

of bone tissue. The genesis of high strength and fracture toughness in bone can be explained in terms of the known concepts embodied in the fracture mechanics of brittle-matrix composites. In contrast, hard dental tissue contains lesser amounts of organic substances, but the mineral component of dentin consists of cylindrical HAp crystals [18].

1.7 Introduction of Teeth: A Natural Biomaterial

Basically, tooth (Figure 1.5) is also a microscopically functionally graded calcium phosphate or HAp-based natural biocomposite material like bone. Furthermore, tooth also has a hierarchical architecture, e.g., from macro-structure to microstructure to nanostructure (Figure 1.5). The tooth comprises mainly the hard enamel, the more ductile dentin, and a soft connective tissue, the dental pulp. Enamel is the hardest structure in the human body, with approximately 96 wt% hydroxyapatite (HAp). On the other hand, dentine possesses a porous structure and is made up of ~ 70% inorganic material

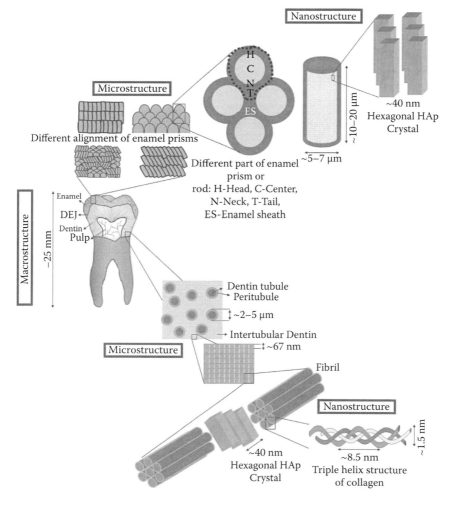

FIGURE 1.5
Schematic of the hierarchical structure of teeth.

(i.e., HAp), ~ 20% organic materials (i.e., collagen fiber), and ~10% water by weight. The enamel microstructure shows different orientations of closely packed enamel prisms or rods. These rods are encapsulated by an organic protein called enamel sheath. Further, the prisms or rods consist of nano-size inorganic HAp crystals with different orientations inside. On the other hand, the dentine has a collagen matrix reinforced with HAp nanocrystals retained in layer by layer. The composite bed of dentine also has dentin tubules, channel-like microstructures that supply the nutrition from the pulp region to the crown part of the teeth. In contrast, the interface between the dentin and enamel junction (DEJ) is arranged with dome-shaped excavations. Therefore, the irregular interface interlocks the two tissues, e.g., enamel and dentine. It is generally seen that the enamel never delaminates from the dentine in spite of millions of repeated masticatory loading during the chewing/grinding of any food, or even during sudden impact loading due to an accident. The microstructure is different in different regions of enamel (Figure 1.6a, b) [21].

1.8 Surface Engineering of Bioinert Materials

Mainly biocompatible metallic materials such as 316 grade stainless steel, Co-Cr-Mo alloy, pure titanium, Ti-6Al-4V, NiTi, etc., which are often used as implants or for prosthesis applications, require special surface engineering because of their poor surface mechanical properties and inferior performance in biological environments. It is well known that the above-mentioned popular biometals are basically bioinert in nature. Thus, in vivo application of these metals could not actively take part in the osseointegration process or bone tissue ingrowth or bone fracture fixation. However, materials like HAp, other CaP materials, and bioglass possess bioactive properties, or in other words, these materials can actively take part in osseointegration. Now a very basic question will rise: Why can we not use these bioactive materials directly for orthopedic application? We cannot because of their poor structural integrity and poor surface mechanical properties. On the other hand, if we look in to metallic materials, their structural integrity is undoubtedly much superior to those of bioactive ceramics and glass. That's why people thought of surface modifications of those biometals and alloys. In two ways it can be useful:

1. Enhancement of biological properties
2. Improvement of surface mechanical properties

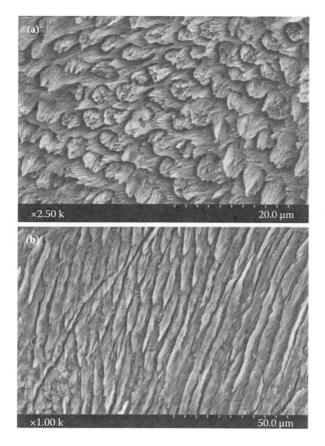

FIGURE 1.6
Enamel rods or prisms within enamel microstructure in different regions: (a) keyhole shaped and (b) longitudinal alignment. (Reprinted/modified from Biswas et al., *Journal of the Institute of Engineers (India): Series D*, 93: 87–95, 2012. From Institute of Engineers (India) and Springer. With permission.)

The first one is basically introduction of HAp or other CaP materials, including bioglass, to increase the osseointegration, bony tissue ingrowth, and environmental shielding of biometals. In general, plasma spaying has been utilized to deposit HAp. The second one is essentially to enhance surface mechanical properties in particular wear and hardness, which is useful in knee implant applications where wear resistance properties of Co-Cr-Mo or Ti alloys should be enhanced. Nowadays, the super-hard TiN coating is deposited on the metallic alloys to increase its hardness and antiwear properties. Alternatively, plasma nitriding method is also popular to enhance the surface mechanical properties of the metallic implants.

1.9 Challenges to Develop Surface-Engineered Implants

For development of surface-engineered metallic implants, the most critical and major requirements are:

1. The presence of excellent adhesion between coating/film and metallic implants not only in an as-grown condition, but also in a simulated or accelerated body environment.
2. The phase purity and crystallinity of deposited bioactive coatings for controlled bioresorption.
3. The adequate porosity level of deposited bioactive coatings for bony tissue ingrowth.
4. The satisfactory bioactivity performance in a simulated or accelerated body environment.
5. The acceptable performance in both short-term and long-term stability tests of the coating.
6. The minimal level of residual stress at the interface to hinder the delamination of the coating.
7. The satisfactory performance of the coatings in the test for metal ion leaching in a simulated or accelerated body environment.
8. The insurance that while bioactive coating deposition is carried out at a high temperature, the compositional and property degradation of the metallic implants is minimal and within acceptable limits as specified by standards.
9. Preservation of a sterile environment while bioactive coating deposition is carried out.
10. The maintenance of uniform coating thickness, particularly in a contoured zone or at the edge of the metallic implants.
11. The strict adherence to mandatory in-depth investigation of the performance of the surface-engineered metallic implants in an animal model prior to human trial.

Thus, we have to be careful when we are going to develop or engineer the surface of a metallic surface for biomedical implantation purposes. One of the major challenges is that more often than not, the metallic implants have complex as well as contoured shapes, and hence are not at all easily amenable to today's available conventional coating deposition technologies practiced for large-scale production. This is an issue of major concern and is therefore of considerable importance. That is why the relative advantages and disadvantages of the methods of bioactive coating deposition will be discussed in Chapter 2.

1.10 Summary

The basic introduction regarding various types of biomaterials has been discussed, with a major emphasis particularly on bioactive materials. The well-known bioactive ceramic material hydroxyapatite and its usage as a bioactive coating has been described. The hierarchical structure of natural hybrid biocomposites incorporating naturally formed hydroxyapatite, e.g., bone and teeth, has been explained. Finally, the need for development of and the challenges involved in surface-engineered biometallic implants has been highlighted.

References

1. Hench L. L. and Ethridge E. C. 1982. *Biomaterials: an interfacial approach*. New York: Academic Press.
2. Williams D. F. 1990. Biocompatibility: an overview. In *Concise encyclopedia of medical and dental materials*, Williams D. F. (Ed.). Oxford: Pergamon Press, 52.
3. Shackelford J. F. 1996. *Introduction to materials science for engineers*. NJ: Prentice Hall.
4. Wintermantel E. and Woo H. S. 1996. *Biokompatible werkstoffe und bauweisen. Implantate für medizin und umwelt*. Berlin: Springer-Verlag.
5. Heimann R. B. 2002. Materials science of crystalline bioceramics: a review of basic properties and applications. *CMU Journal*, 1: 23–46.
6. Shackelford J. F. 2005. *Bioceramics*, vol. 1. Amsterdam, The Netherlands: Gordon and Breach Science Publishers.
7. LeGeros R. Z. and LeGeros J. P. 1997. Bone substitute materials and their properties. In *Knochenersatzmaterialien und wachstumsfaktoren*, Schnettler R. und Markgraf E. (Eds.). Stuttgart: Thieme, 12–18.
8. Hench L. L. 1993. Bioceramics: from concept to clinic. *American Ceramic Society Bulletin*, 72: 93–98.
9. Hench L. L., Splinter R. J., Allen W. C., and Greenlee T. K. 1971. Bonding mechanisms at the interface of ceramic prosthetic materials. *Journal of Biomedical Materials Research*, 5: 117–141.
10. Martin R. B. 1996. Biomaterials. In *Introduction to bioengineering*, Berger S. A., Goldsmith W., and Lewis E. R. (Eds.). Oxford: Oxford University Press, 339–360.
11. Narayanan R., Seshadri S. K., Kwon T. Y., and Kim K. H. 2008. Calcium phosphate-based coatings on titanium and its alloys. *Journal of Biomedical Materials Research B*, 85: 279–299.
12. Kumta P. N., Sfeir C., Lee D. H., Olton D., and Choi D. 2005. Nanostructured calcium phosphates for biomedical applications: novel synthesis and characterization. *Acta Biomaterialia*, 1: 65–83.
13. Soballe K. 1993. Hydroxyapatite ceramics for bone implant fixation. Mechanical and histological studies in dogs. *Acta Orthopaedica Scandinavica Supplementum*, 255: 1–58.

14. Hench L. L. 1991. Bioceramics: from concept to clinic. *Journal of the American Ceramic Society*, 74: 1487–1510.
15. Katz J. L. 1980. *The mechanical properties of biological materials*. Cambridge: Cambridge University Press.
16. Orlovskii V. P., Komlev V. S., and Barinov S. M. 2002. Hydroxyapatite and hydroxyapatite-based ceramics. *Inorganic Materials*, 38: 973–984.
17. Suchanek W. and Yoshimura M. 1998. Processing and properties of hydroxyapatite-based biomaterials for use as hard tissue replacement implants. *Journal of Materials Research*, 13: 94–117.
18. Liu H. and Webster T. J. 2007. Bioinspired nanocomposites for orthopedic applications. In *Nanotechnology for the regeneration of hard and soft tissues*, Webster T. J. (Ed.). Singapore: World Scientific Publishing.
19. Aoki H. 1991. *Science and medical applications of hydroxyapatite*. Tokyo: Takayama Press System Centre, 165.
20. Martin R. B. 1999. Bone as a ceramic composite material. *Materials Science Forum*, 7: 5–16.
21. Biswas N., Dey A., Kundu S., Chakraborty H., and Mukhopadhyay A. K. 2012. Orientational effect in nanohardness of functionally graded microstructure in enamel. *Journal of the Institute of Engineers (India): Series D*, 93: 87–95.

2

Plasma Spraying and Other Related Coating Techniques

Here, we shall discuss the basics and history of plasma spraying in general and microplasma spraying in particular. The method of hydroxyapatite (HAp) coating deposition by different techniques and the relative merits and demerits of those will be also described. The difference between conventional atmospheric high-power macroplasma and microplasma will be highlighted to demonstrate that such differences can and do indeed lead to significant changes in the correspondingly deposited HAp structures.

2.1 Plasma Spray Process

The conventional atmospheric plasma spraying or macroplasma spraying (MAPS) process is relatively straightforward in concept. However, it is rather complex in function. The plasma gun operates on direct current. The plasmatron actually provides power to the plasma gun. It is the heart of the plasma spraying machine. The efficacy of a given MAPS process and the resultant coating development are strongly sensitive to the plasmatron power. However, as far as the chemical quality, physical quality, microstructure, phase purity, etc., of a coating are concerned, the plasmatron power is not the sole determining factor, although it plays a major role. The current that the power of the plasmatron source provides to the plasma gun helps to sustain a stable nontransferred electric arc between a tungsten cathode and a copper anode. The annular-shaped anode is usually water cooled. A plasma gas is introduced at the back of the gun interior. Generally, the plasma gas is argon or nitrogen or any other inert gas mixed typically with less than 1–5% hydrogen to enhance enthalpy. The gas swirls in a vortex and finally gets out of the front exit of the anode nozzle. The electric arc from the cathode to the anode completes the circuit. As a result, an exiting plasma flame generally forms on the outer face of the latter. The plasma flame axially rotates due to the vortex momentum of the plasma gas. For a typical DC plasma torch operating at 10–40 kW, the temperature of the plasma just outside of the nozzle exit is effectively in excess of 15,000 K. However, for many commercial applications, the plasma torch may operate at a power level that

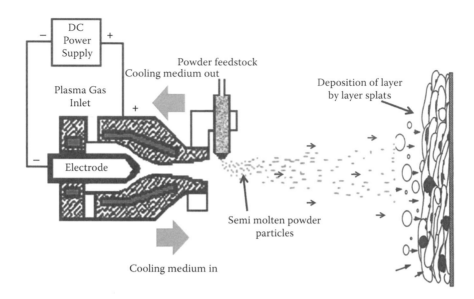

FIGURE 2.1
Schematic of a typical atmospheric plasma spray or macroplasma (MAPS) process.

could be even higher than 100 kW. The temperature of the plasma drops off rapidly as one traverses away from the exit of the anode. This is why the powder to be processed is introduced at this hottest part of the flame. The size of the powder particles to be used in the plasma spray process is mainly governed by the nozzle size of the powder chamber and the net plasmatron power. During the plasma spraying process, the powder particles are accelerated and melted in the flame on their high-speed (~100–300 m.s^{-1}) path to the substrate. Subsequently, they impinge upon the substrate and undergo rapid solidification. A schematic of a typical plasma spray process is shown in Figure 2.1. This process can also be carried out in a vacuum environment to reduce the chance of oxidation of the substrate at higher temperature.

2.2 How Will Coating Form?

Generally, the particles experience melting during the plasma spraying processes. However, some large particles may not be completely molten, as they stay for only a short duration in the plasma arc zone. The droplets are propelled and finally impacted onto the substrate surface, forming splats. Individual splats assume the shape of a newly formed pancake. As the sprayed particles impinge upon the substrate surface, they cool and build up, splat by splat, into a lamellar architecture, thus developing the

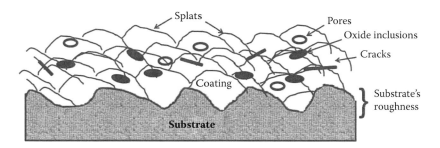

FIGURE 2.2
Schematic of the cross sectional view of a typical plasma sprayed coating.

microstructure of the plasma sprayed coating. The coating thus formed is not homogeneous and typically contains a certain degree of porosity. In fact, in the case of plasma sprayed metals, the coating can even contain oxides of the metal. Figure 2.2 schematically depicts the cross section of the lamellar structure for a typical plasma sprayed coating that also contains oxide inclusions, cracks, and pores.

2.3 Plasma Sprayed HAp Coatings

The rapid dissipation of heat to the substrate makes the splats solidify quickly. Such a rapid solidification process often leads to formation of some metastable phase/phases in the coating. Thus, in general, plasma sprayed coatings may well possess nonstoichiometric phases and an inhomogeneous microstructure. In case of a plasma sprayed HAp coating, therefore, both amorphous calcium phosphates and crystalline HAp may occur in the coating [1]. The HAp of these coatings has a lower Ca/P ratio than the starting HAp material. This deviation happens because the powder reaches very high temperatures during deposition [2, 3]. Further, the rapid solidification during the plasma spraying process often forms the amorphous calcium phosphate phase. The reason is simply that it might not have had enough time to crystallize within a short duration of flight time at high temperatures [2, 3].

2.4 Microplasma Spraying

The MIPS process is similar to the MAPS process, as shown in Figure 1.1, except the range of input power and miniaturization of machines. Similar to the MAPS process, here also an arc is produced between a tungsten cathode and a copper anode, and it transfers to a continuous current path. Due to high potential difference, a pilot arc is initiated between the tungsten cathode and the continuously water cooled copper anode. Primary gas, i.e., plasma gas (Ar), is passed between the cathode and anode. The presence of the pilot arc ionizes the argon gas. Now this ionized argon gas provides a continuous current path for the main transferred plasma arc. In our own experiments, argon was also used as the secondary gas, i.e., the shielding gas, to protect the metallic substrate from oxidation. Hot and partially molten externally fed powder was flown by the Ar gas pressure, and it stuck layer by layer on the substrate in the form of splats. As mentioned earlier, these splats also assume the shape of a newly baked pancake. Subsequently, the splats were air cooled for the development of coating. The present authors utilized a commercial MIPS machine (Miller Maxstar 200 SD 2.5 kW, USA) at a low plasmatron power of <1.5 kW with an external powder feeder chamber. The input current was fixed at 40 A for the optimized HAp coating. A water cooling unit was used to cool the copper anode arrangement. The details of optimized process parameters are given in Table 2.1. It is of paramount importance to optimize the process parameters to obtain reproducible coating quality and microstructure. Keeping this particular aspect in view, as a typical illustrative example, the optimization of process parameters for MIPS HAp coating development has been discussed in detail later in this chapter.

Figure 2.3a–e gives a bird's-eye view of the portable MIPS machine setup utilized in the experiments conducted by the present authors. The electrical power unit is depicted in Figure 2.3a. Figure 2.3b provides a view of the Ar gas chamber. A picture of the water cooling unit is presented in Figure 2.3c.

TABLE 2.1

Optimized Process Parameters for HAp Coating

Parameters	Values
Primary gas pressure (Ar)	4 bar at 20°C
Secondary gas pressure (Ar)	4 bar at 20°C
Primary gas (plasma gas) flow rate	10 SLPM
Secondary gas (shielding gas) flow rate	20 SLPM
Powder deposition rate	~1.5 mg/s
Powder size	~−53 + 64 μm
Input current	~40 amp
Input voltage	~30 V
Plasmatron power	1.2 kW
Stand-off distance	75 mm
Rotational speed of sample	150 rpm
Distance between cathode and plasma nozzle	1.7 mm
Distance between anode and plasma nozzle	1 mm

Note: SLPM = standard liters per minute.

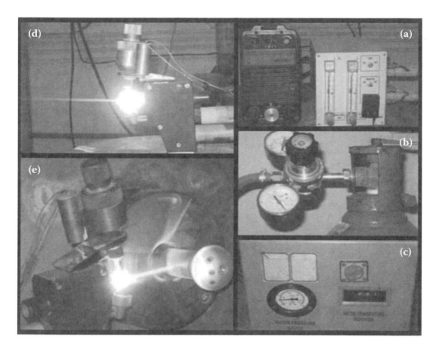

FIGURE 2.3 (See color insert.)
Microplasma spraying setup at CSIR-CGCRI, Kolkata: (a) electrical power unit, (b) primary (Ar) and secondary gas (Ar) chamber, (c) cooling unit, (d) plasma gun setup with external powder feeder, and (e) MIPS-HAp coating process in progress with a rotating holder.

Finally, a view of the plasma gun set up with an external powder feeder is shown in Figure 2.3d. MIPS was carried out on a custom-built SS316L metallic shell for hip implant, mounted on a rotating holder with an external powder feeder chamber (Figure 2.3e).

2.5 Microplasma Spraying and Its Application

The microplasma spraying technique was introduced in the mid-1990s, and researchers started reporting the work in the late 1990s and early 2000s. Rather than commercialization, in-house custom-built faculties of MIPS were popular. The universities and research establishments like Xi'an Jiaotong University, Dalian Maritime University, Stanford University, Paton Welding Institute, Xi'an Hi Tech Research Institute, RWTH Aachen, etc., developed in-house MIPS facilities. The history of the MIPS process is summarized in Table 2.2 [4–41]. Metals, alloys, metal oxides, bioactive ceramic, composite powder, etc., can be deposited by the MIPS facility. The data reflected in

TABLE 2.2

History of MIPS Process

Year	M/c Details	Coating	Plasma Gas/ Primary Gas	Carrier Gas/ Powder Feed Gas/Secondary Gas	Arc Current (A)	Power (kW)	Spray Distance (mm)	Reference
1994	MIPS	α and γ Al$_2$O$_3$	—	—	—	—	—	[4]
1998	MIPS	Fe, Ni, Cu alloys	—	—	—	—	30–50	[5]
1999	MIPS	Ni alloy	Air + 10% propane-butane	—	40–45	1–1.3	80	[6]
1999	In-house developed: Xi'an Jiaotong University	NiCrBSi	Argon	Argon or nitrogen	10–80	<2.4	20–120	[7]
2002	Plasma-LE15, in-house developed: Dalian Maritime University	Al$_2$O$_3$	Argon	Hydrogen + nitrogen	50–80	2.5–4	40	[8]
2003	MIPS	Al$_2$O$_3$	Argon	Argon	—	2.2, 3.9	10	[9]
2003, 2004	In-house developed: Xi'an Jiaotong University	Cu	Argon	Argon	80–120	2.8–4.2	40–100	[10, 11]
2005	Vacuum MIPS, in-house developed: Stanford University	SS316L, Ti-6Al-4V, W	Argon + hydrogen	—	—	1–3	—	[12]
2006	MPN-3, Paton Welding Institute, Ukraine	HAp	Argon	Argon	35–50	—	50–80	[13]

Year	System	Material	Gas	Gas				Ref.
2006	MPN-3, Paton Welding Institute, Ukraine	Ni-20Cr, Al$_2$O$_3$/TiO$_2$	—	—	32–40	—	50	[14]
2006	Modified MIPS	Metal oxides	Argon	—	—	—	—	[15, 16]
2006	MIPS	MoO$_2$, MoO$_3$	Argon	—	—	—	—	[17]
2007	MPN-3, Paton Welding Institute, Ukraine	HAp	—	—	30–40	2	50–140	[18]
2007	MIPS	Al$_2$O$_3$-13 wt% TiO$_2$ (AT13)	Argon	Nitrogen	100–150	—	40	[19]
2007	Multifunction microplasma spraying system	Cu	Argon	Hydrogen	—	—	40	[20]
2008	MIPS: Xi'an Hi Tech Research Institute, China	Al2O3-13 wt% TiO$_2$ (AT13)	Argon	—	—	—	—	[21]
2008	MIPS: Xi'an Hi Tech Research Institute, China	Al$_2$O$_3$	Argon	—	—	—	—	[22]
2011	MIPS	CNTs-SiC/Al$_2$O$_3$-TiO$_2$	Argon + nitrogen	Nitrogen	250	—	100	[23]
2011	MIPS	HAp	—	—	—	—	—	[24]
2010, 2011	MIPS facility, RWTH Aachen, Aachen, Germany	HAp	—	—	—	—	—	[25–27]
2011	MIPS	Al$_2$O$_3$-TiO$_2$	—	—	250	—	—	[28]
2011	Tabletop MIPS, Spraymet, Bangalore, India	HAp, bioglass	—	—	—	—	—	[29]
2009–2014	MIPS: Miller Maxstar 200 SD 2.5 kW, USA	HAp, TCP	Argon	Argon	30–40	<1.5	50–100	[30–41]

Table 2.2 show that in almost all the cases, the primary or carrier gas utilized was argon. There were also instances, however, where the utilization of a mixture of other inert gasses had been reported. The plasmatron power was reported to be widely varying, e.g., in the range of 1–4.2 kW. The corresponding arc current had a range as wide as 10–250 A. As the plasmatron power utilized in a MIPS process is much less than that utilized in a MAPS process, it is obvious that the spraying distance will have to be much smaller than what could be afforded for a MAPS process. Indeed, the spraying distance was much less, e.g., in the range of 10–140 mm. The statistical width of the range reflects the variations in the corresponding plasmatron power (1–4.2 kW) and the consequent gun current (10–250 A). The data presented in Table 2.2 also highlight that since 2006, the usage of the MIPS process to develop HAp coatings has been gaining in popularity. Very recent experimental results published by our own research group [30–41], as well as other research groups [25–27] in animal models, have proved beyond doubt the superior efficacy of MIPS-HAp-coated biomedical implants.

2.6 Microplasma Spraying: A Unique Manufacturing Technique

Now the reader may genuinely seek an explanation to the question of what is the unique reason that makes researchers interested in depositing both metallic and ceramic coatings by the MIPS process, which characteristically utilizes a plasmatron power that is much lower than that of a conventional MAPS process. To provide a rational explanation to such genuine queries, we ought to take a critical look at the unique advantages of the MIPS process. The MIPS process uses low plasmatron power. So the heat content is less. Therefore, one of the unique advantages of the MIPS process is that it produces an amount of thermal and residual stresses that is much lower than what would be provided by a conventional MAPS process. There are many biomedical and industrial applications that require the coating of implants and machine parts of very tiny sizes. A major advantage of the MIPS process is that it has a spot size much smaller than what could be afforded by the conventional MAPS process. Hence, development of plasma sprayed coatings on tiny sizes can be done by the MIPS process with efficacy much superior to that of the usual MAPS technique. One particular case in sight is that of the small-area defect generation in continuously run gas turbine blades made of very costly super alloys. Because of the small-area defect, the blade cannot be thrown away. At the same time, the turbine blade cannot be utilized in the presence of the defect. It is in such utter situations that the MIPS process came out very handy with its small spot size, which could be utilized very effectively and efficiently to deposit the coating of desired composition only on the small-area defect, thereby providing a very useful means of quick, precise, localized repair that paved the way for further repeated usage of the affected gas turbine blades. It is interesting to note that as early as 2006, researchers in this area recognized this potential of the MIPS process during the developmental stages of this technology, as reflected in the data presented in Table 2.3 [42–57]. In fact, the data contained in Table 2.3 summarize the key features of some of the important intellectual properties (IPs) regarding utilization of the MIPS process. That is why, perhaps, it is not surprising to notice from the data located in Table 2.3 that since 2006, there have been several IPs granted to both US and European researchers. This vital information only reinforces the huge potential of the MIPS process

TABLE 2.3

Patents Related to Microplasma Spraying

Year of Grant	Country/Region	Salient Features of the Invention	Reference
2006	Europe	Method for repairing a workpiece using microplasma having a width of about 0.5–5 mm without masking the workpiece with one or more number of powder	[42]
2006	Europe	A method and apparatus for repairing defective areas of the thermal barrier coating in gas turbine engines; can be patched on the component without masking the component	[43]
2006	United States	Portable microplasma spray apparatus can be transported to on-site locations to facilitate quick repair work	[44]
2006	United States	Method and apparatus for plasma flame spray for copper-nickel, aluminum-copper, and other similar composition materials introducing coating material into the plasma effluent for gas turbine application	[45]
2006	United States	Method and apparatus for repairing a thermal barrier coating on components in gas turbine engines with microplasma spraying	[46]
2007	United States	A method for repairing a workpiece using microplasma; includes a microplasma stream having a width of about 0.5–5 mm without masking the workpiece	[47]
2007	Europe	Microplasma spray apparatus for boron- and nickel-based combination powder deposition	[48]
2008	Europe	Portable, handheld microplasma spray coating apparatus	[49]

continued

TABLE 2.3 (continued)

Patents Related to Microplasma Spraying

Year of Grant	Country/Region	Salient Features of the Invention	Reference
2009	Europe	Microplasma (0.5–4 kW) spray nickel chrome alloy and a nickel chrome-chrome carbide alloy coating apparatus with argon as plasma gas for repairing turbine vane	[50]
2009	Europe	Same as EP1652951 B1 [50] for coating of copper, nickel alloy, and aluminum-copper alloy	[51]
2009	USA	Microplasma apparatus for repairing workpiece comprising a surface having a localized damage site	[52]
2009	USA	A method and apparatus for microplasma spray coating a portion of a turbine vane without masking any portions	[53]
2010	USA	A method and apparatus for microplasma spray coating a portion of a substrate, such as a gas turbine compressor blade, without masking any portions	[54]
2010	USA	A method to coat a workpiece such as a gas turbine compressor blade	[55]
2011	USA	Microplasma spray apparatus for boron- and nickel-based combination powder deposition for repairing gas turbine blades without masking	[56]
2011	USA	A method for repairing an area of damaged coating in a component of a gas turbine engine by means of microplasma spraying	[57]

in the biomedical as well as industrial application areas, where very quick, precise, small-area coating depositions are required.

2.7 Other Coating Processes

Besides conventional and commercially accepted plasma spraying methods, there are many other methods to develop HAp coatings. These typically include by way of references, but may not be strictly limited to, methods such as high-velocity oxy fuel (HVOF) [58–60], pulsed laser deposition (PLD) [61–66], cold spraying [67, 68], liquid precursor plasma spraying (LPPS) [69, 70], electrophoretic deposition (EPD) [71, 72], electrohydrodynamic atomization (EHDA) [73, 74], sol-gel and dip coating [75–81], biomimetic [82–84], and

TABLE 2.4

Relative Merits and Demerits of HAp Coatings Developed by Different Processes

Process or Technique	Thickness	Porosity	Crystallinity	Phase Purity	Microstructure	Adhesion Strength[a]	Remarks
MIPS including vacuum environment	50–300 μm	10–20% or more	80–92%	Almost phase pure	Microcracks	Moderate	• High deposition rate • Moderate cost
MAPS including vacuum environment	50–300 μm	Dense/up to 18%	<55%	Amorphous/impure phases	Microcracks/macrocracks	Good	• High deposition rate • Moderate cost • Chance of residual stress • Formation of amorphous and impure phases • Maintenance of uniformity of thickness in contours difficult • Coating on polymeric biomaterials not possible
HVOF	~200 μm	Dense	Low	Amorphous/impure phases	Microcracks/macrocracks	Good	• Same as MAPS
PLD	0.05–5 μm	Dense	Low	Amorphous/impure phases	—	Good	• High cost
Cold spray	20–30 μm	Dense	—	Almost phase pure		—	• Ceramic coating is difficult due to its having less ductility

continued

TABLE 2.4 (continued)

Relative Merits and Demerits of HAp Coatings Developed by Different Processes

Process or Technique	Thickness	Porosity	Crystallinity	Phase Purity	Microstructure	Adhesion Strength[a]	Remarks
LPPS	—	Dense/up to 48%	—	Phase pure		—	• High cost and complex process • Both dense and porous microstructure can be achieved • Nanostructured coating can be produced
EPD	<10 µm	Dense	Good	Almost phase pure	Microcracks/macrocracks	—	• Low cost • High deposition rate • Uniform coating • Coating possible on complex-shaped substrates • Difficult to avoid formation of crack while post-annealing
EHDA	—	Dense	—	Almost phase pure	—	—	• Simple processing method • Cost-efficient • Nanostructured coating can be deposited • Patterning can be done • Compatibility with microfabrication technology

Sputtering	0.5–3 μm	Dense	—	Amorphous	Uniform	Good	• Low deposition rate but uniform coating • High cost • Process complexity high • Coating possible on only flat surface of metal or nonmetals
Sol-gel and dip coating	3–500 μm	Dense/nanoporous	—	Almost phase pure	Microcracks/macrocracks	Moderate/good	• Inexpensive • Uniform coating achievement is difficult • Simple process • Post-annealing required while often cracks formed due to thermal coefficient mismatch between coating and substrate • only on flat surface of metal or nonmetals • Coat on contour- and complex-shaped substrate of metals and nonmetals is possible
Biomimetic	<30 μm	Dense/nanoporous	—	Phase pure	—	Low-moderate-good	• Time-consuming process • Bone-like apatite formation that may further assist bone fracture healing

Notes: Bonding strength value: Low < 10 MPa, moderate = 10–15 MPa, good > 15 MPa. MIPS = microplasma spraying, MAPS = macroplasma spraying, PLD = pulsed laser deposition, HVOF = high-velocity oxy fuel, EHDA = electrohydrodynamic atomization, EPD = electrophoretic deposition, LPPS = liquid precursor plasma spraying.

sputtering [85, 86]. Now briefly, the aforesaid techniques will be described as follows. Based on existing literature, an exhaustive survey of the comparative merits, demerits, and features of all the processes involved in the deposition of HAp coatings has been summarized in Table 2.4. The data presented in Table 2.4 suggest that if we consider properties like crystallinity, phase purity, porosity, bonding strength, process simplicity, cost-effectiveness, and the acceptably high rate of deposition of the deposited coatings, the MAPS process in general and the MIPS process in particular emerge as the leading contenders of the top position from among those reported so far. There is no doubt that these relative advantages, as reflected in the data presented in Table 2.4, have remained the major catalyst that basically promoted their commercialization.

2.7.1 High-Velocity Oxy Fuel

The high-velocity oxy fuel (HVOF) process is basically a modified version of thermal spray technology developed in 1980. The theory of the HVOF process is based on combustion of a gaseous or liquid fuel and oxygen in a combustion chamber. They react with each other in the chamber to form hot gas. Examples of gaseous fuel are, e.g., hydrogen, propylene, propane, acetylene, natural gas, etc., while kerosene is a typical example of a liquid fuel. The combustion process generates a huge supersonic flame. The pressure generated during the combustion process can reach as high as 1 MPa, and the generated hot gas can have a combustive wave front speed of more than 1000 $m.s^{-1}$. This combination of high pressure and speed is reflected in the flame that propagates at supersonic speed and ultimately propels the powders or particles exposed to the flame at a speed of about 700–800 $m.s^{-1}$ on to the substrate. The deposited coating will be dense in nature. However, it is almost a characteristic feature of coatings produced by this process that due to differences of the thermal coefficients of expansion between the substrate and the coating material, the microcracks will form more often than not, and a significant amount of residual stress can remain frozen in the coatings developed by this process.

2.7.2 Pulsed Laser Deposition

Pulsed laser deposition (PLD) is a basically a potential thin-film deposition technique in a wide category of physical vapor deposition (PVD). In PLD, inside a vacuum chamber, a pulsed laser high-energy beam is focused to strike a target material that is to be deposited. It is well known that the process will be always complex, as high vacuum is involved. Due to high temperature, the target material will be vaporized and form a plasma plume. The vaporized material will be deposited on a substrate that is suitably placed inside the chamber. PLD can be processed in a ultra-high vacuum condition or in the presence of a reactive gas like oxygen. There are many basic physical

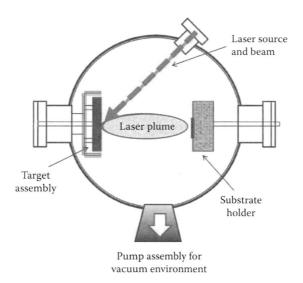

FIGURE 2.4 (See color insert.)
Schematic representation of PLD process.

steps involved in such a process technology. At first, of course, the energy from the incident laser pulse is absorbed by the target. This absorbed energy is at the first stage converted to electronic excitation. Later, the amount of energy absorbed is again released back as the electrons fall back to their ground states from the temporarily excited states. This secondarily released burst of energy is then shared by the target material atoms in three major excitation components, e.g., thermal, chemical, and mechanical. This secondary storage of energy finally results in several processes, such as evaporation, ablation, and plasma formation, that end up in coating deposition. Not necessarily always, but occasionally even localized exfoliation of the target materials surface can also happen as a result of exposure to an incidence of pulsed laser on it. One major limitation of this process, however, is that utilizing it, the growth of thick, e.g., ~100–200 µm films, is very difficult, if not impossible. The schematic representation of PLD is shown in Figure 2.4.

2.7.3 Cold Spraying

The cold spraying technique is basically the advanced form of thermal spraying. In recent days, it has been gaining popularity due to its ability to deposit coatings at a temperature that is much lower than that of the thermal spray process. While the high temperature of conventional thermal spraying provides a scope to deposit dense coatings, it is also the main cause behind microcracking and frozen-in residual stress, which could be detrimental for in-service reliability of such coatings. In the cold spraying technique, therefore, the arrangement is made that instead of temperature, the very high

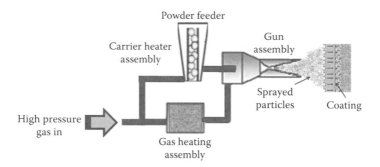

FIGURE 2.5 (See color insert.)
Schematic representation of cold spraying process.

velocity of the gas jet assists to deposit powder particles on the substrate. A typical schematic representation of the cold spray technique is shown in Figure 2.5. The details of the cold spraying technique will be discussed in Chapter 3.

2.7.4 Liquid Precursor Plasma Spraying

The liquid precursor plasma spraying (LPPS) technique is also an emerging method of plasma spray technique to develop nanostructured coating, where coating materials can be deposited from its slurry or liquid precursor. This step is followed by deposition of the powder particle by the conventional plasma spraying technique. Further details of the LPPS process will be discussed in Chapter 3.

2.7.5 Electrophoretic Deposition

The electrophoretic deposition (EPD) technique basically involves a guided relative motion of particles suspended in a fluid. Here, the guide is a spatially uniform electric field that is applied externally. It is well known that the particles can be made to remain suspended in a given fluid. This happens due to the electric charge on the surfaces of the particles. It follows easily that all such surface charges are screened by an ion layer that bears the same absolute magnitude of charge but has an electrical character just opposite to that of the surface charge. This is what maintains the electrical charge balance according to the well-established double-layer theory. What is done in the EPD technique is that an external electric field is deliberately applied to perturb this electrical charge balance. This field provides an additional Coulomb force of electrostatic origin. As a result of this, a net electrical guiding force is generated. This net force guides the motion of the ions in the ion layer onto the cathode, where the coating deposition finally takes place. Although simplistically described, there are many interesting yet complex interfacial

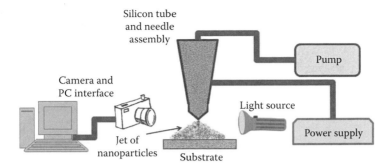

FIGURE 2.6 (See color insert.)
Schematic representation of EHDA process.

physical and chemical processes involved in the EPD technique, which may well deserve separate, dedicated discussion that is beyond both the perspective and scope of the present work. Nevertheless, it is worth mentioning that this coating deposition technique can provide better process control. The deposited coatings possess good crystallinity. But here, again, the major limitation of this technique is that the coating thickness is limited to ~10 µm.

2.7.6 Electrohydrodynamic Atomization

The electrohydrodynamic atomization (EHDA) method is an emerging method to produce thin nanostructured films/coatings as well as predesignated patterns on a given substrate surface [73, 74, 87]. In this process a liquid suspension is used. This suspension consists of the nanoparticles of the material to be coated. The basic principle of this process is the dissociation of the liquid suspension into tiny, atomized spray droplets under the application of an electrical field. Thus, as schematically depicted in Figure 2.6, the suspension that has to be coated is jetted from a needle under application of an electric field in the EHDA process. However, there are two major limitations of the coatings developed by the EHDA process. The first one is that it is very difficult to develop thicker coatings by this process. The second one is that for the coatings deposited by the EHDA technique, the adhesion of the coating to the substrate often remains far from satisfactory.

2.7.7 Sputtering

Today's emerging scenario is such that not only for the applications involving the electronics industry, but also for the applications involving the tribological, biomedical, and all other relevant industrial areas where films may be needed, the sputtering technique is receiving an ever-increasing level of interest. The sputtering technology today has become the most versatile technique in thin-film technology for preparing thin solid films of almost any material in the periodic table. The sputtering method is one of the most well-known, proven physical vapor deposition (PVD) techniques. It is the only method that has become a serious competitor to the better established vacuum evaporation (VE) technique. The VE technique involves evaporation of the target by thermal heating in vacuum condition and sublimation/deposition of the same on a desired substrate kept inside the deposition chamber. In the sputtering process, on the other hand, the material to be deposited is called the target. It is connected to a negative voltage supply. So, it works as the cathode. The substrate on which the coating is to be deposited is placed facing the cathode. It is connected to the positive voltage supply. So it works as the anode. Now, to start with, a controlled gas, for instance, chemically inert argon, is introduced into a vacuum chamber. The chamber is typically evacuated up to 10^{-6} to 10^{-7} mbar with the combination of roughing, diffusion, and turbo molecular pumps. Then the cathode is electrically energized to generate a self-sustaining plasma. The gas atoms become positively charged ions by losing electrons within the plasma. Therefore, these positively charged argon ions are accelerated with adequately high kinetic energy to hit the target that acts as the cathode. This high-energy bombardment results in the ejection or sputtering out of the neutral atoms from the target surface along with the charged atoms and electrons. It is these neutral atoms sputtered out of the target surface that condense into thin films on the substrate that works as the anode. Thus, we can understand the physical picture of the sputtering process in a more elaborate manner in the following way. To start with, a controlled gas, for instance, chemically inert argon, is introduced into a vacuum chamber. The chamber is typically evacuated up to 10^{-6} to 10^{-7} mbar with the combination of roughing, diffusion, and turbo molecular pumps. Then the cathode is electrically energized to generate a self-sustaining plasma. As mentioned earlier, the exposed surface of the cathode, referred to as the target, is a piece of the material to be deposited over the substrates. The gas atoms become positively charged ions by losing electrons within the plasma. Therefore, these positively charged argon ions are accelerated with adequate kinetic energy to hit the target. This bombardment sputters out atoms or molecules from the target material. The sputtered-out material now consists of a vapor stream, comprised of neutral atoms along with charged atoms and electrons. It is the neutral atoms of this vapor stream that pass through the chamber, and strike and stick to the substrate as a film or coating. The surface of the substrate needs to be kept clean so as to obtain good film adhesion. Suitable cleaning and handling procedures must be used before placing substrates into the vacuum chamber. Moreover, it is common to incorporate in situ cleaning features like sputter etch into the sputter system as an option. Various measures may have to be taken in order to obtain the desired film properties. The sputter system design allows process engineers to adjust a number of parameters to obtain desired results for variables, such as grain structure, uniformity, thickness, stress, adhesion strength, optical or electrical properties, and much more. The type of power

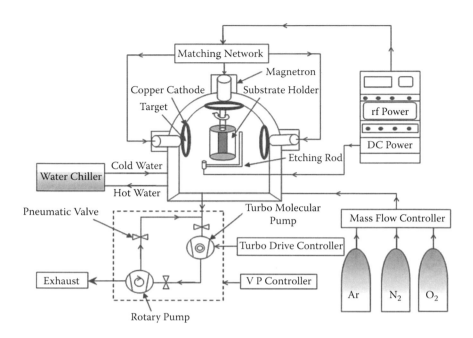

FIGURE 2.7
Schematic representation of a typical three-target rf magnetron sputtering process. (Drawing courtesy of Ms. Deeksha Porwal.)

to be employed on the cathodes must also be taken into account. For instance, a DC power is ideal for conductive materials. The corresponding sputtering technique is known as DC or direct sputtering technique. On the other hand, radio frequency (rf) power is also capable of sputtering nonconductive materials, e.g., ceramics. The corresponding sputtering technique is known as the radio frequency sputtering technique. Pulsed DC is suitable for some techniques, such as reactive sputtering. What happens in the reactive sputtering process is that it employs the combination of a reactive gas, e.g., oxygen, and an elemental target material, i.e., silicon. Therefore, the reactive gas, i.e., oxygen, chemically reacts with the sputtered atoms within the chamber. This reaction leads to generation of a new compound. It is this newly formed compound that serves as the coating material, rather than the original pure target material. As a thin-film or coating deposition technique, the major advantages of the sputtering technologies include, but are not necessarily limited to, good adhesion, high uniformity of film thickness, maintenance of the stoichiometry of the original target composition, and epitaxial film growth possibility at low temperature. However, there are two major limitations of this technology as well. One major disadvantage is that usually the film deposition rate is low. The other problem is that the target material must be in sheet form. A schematic representation of a typical experimental setup for rf magnetron sputtering with three targets is shown in Figure 2.7.

2.7.8 Sol-Gel and Dip Coating

We can define the sol-gel technique as a process in which the formation of an oxide network happens through polycondensation reactions of a molecular precursor in a liquid. It is mainly due to the need of new synthesis methods in the nuclear industry that the genesis of this process actually dates back to 1960s. Thus, in a simplistic manner, we can try to understand the basic idea behind sol-gel synthesis. Here, the idea is to "dissolve" the compound in a liquid in order to bring it back as a solid in a controlled manner. One of the biggest advantages of this is method is that it is a low-temperature process. It enables mixing at an atomic level. It can give uniformly sized small and even nanosized particles, which can be sintered easily later on by conventional processing techniques. Moreover, it can prevent the problems with co-precipitation, which may be inhomogeneous, being a gelation reaction. Further, multicomponent compounds may be prepared with a controlled stoichiometry by mixing sols of different compounds. The best part of the technique is that the precursor sol can be deposited on a substrate to form a film. The most popular method of achieving the same is by dip coating or spin coating. Before we can discuss this technique further, it is imperative to have some basic idea about what a sol is and what a gel is. First, let us understand what a sol is. Simplistically speaking, a sol is a stable colloidal suspension. It is such a suspension that has a heterogeneous structure, can scatter light, and hence is capable of exhibiting the well-known Tyndall effect. Examples of naturally occurring sols are milk, blood, smoke, etc. A solution, on the other hand, is stable but homogeneous in structure, cannot scatter light, and hence is incapable of exhibiting the well-known Tyndall effect. Thus, it may be understood that a sol is a stable dispersion of colloidal particles or polymers in a solvent. Here, the particles may be either crystalline or amorphous. The major difference between a sol and an aerosol is that a sol contains particles in a liquid medium, while the aerosol contains particles in a gaseous medium.

Now let us try to understand what a gel is. A gel, as the name suggests, is a jelly-like structure. Actually, it consists of a three-dimensional continuous network. This continuous network encloses a liquid phase. The question that faces us now is that how the network is formed. Well, if it is a colloidal gel, the agglomeration of the colloidal particles frame the three-dimensional network. What happens in a polymeric gel will of course be different from what happens in a colloidal gel. In a polymeric gel, the particles form a polymeric substructure. This becomes possible by the aggregation of the subcolloidal particles. Highly localized extensive linking of polymer chains can also lead to gel formation. It is to be kept in mind, though, that in most of the situations involving material synthesis, the nearest neighbor interactions are of a covalent nature, while in a sol the suspended nearest neighbors or particles can and do mostly interact through forces that belong to either van der Waals forces or hydrogen bonds.

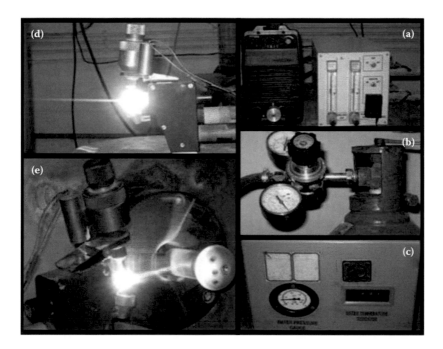

COLOR FIGURE 2.3
Microplasma spraying setup at CSIR-CGCRI, Kolkata: (a) electrical power unit, (b) primary (Ar) and secondary gas (Ar) chamber, (c) cooling unit, (d) plasma gun setup with external powder feeder, and (e) MIPS-HAp coating process in progress with a rotating holder.

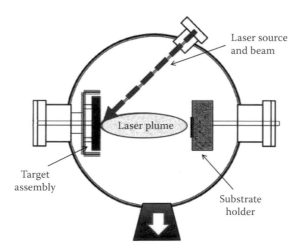

COLOR FIGURE 2.4
Schematic representation of PLD process.

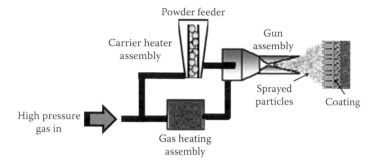

COLOR FIGURE 2.5
Schematic representation of cold spray process.

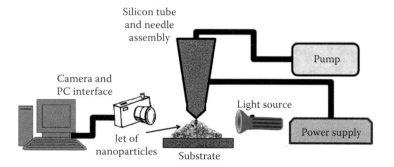

COLOR FIGURE 2.6
Schematic representation of EHDA process.

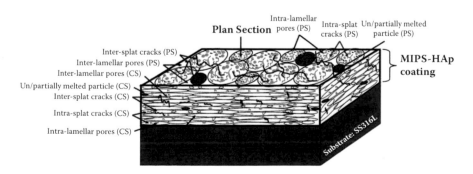

COLOR FIGURE 4.8
Schematic of the structure and nature of the MIPS-HAp coating. (Reprinted from Dey et al., *Journal of Materials Science*, 44: 4911–4918, 2009. With permission from Springer.)

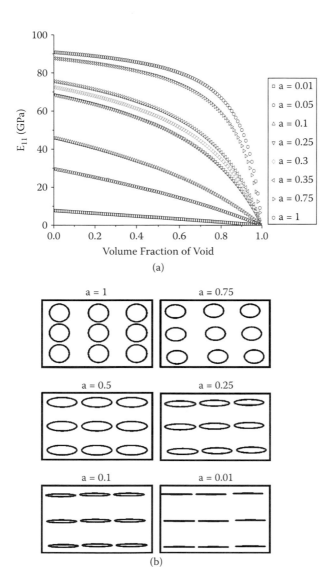

COLOR FIGURE 4.20
(a) Predicated variation of E_{11} as a function of the volume fraction of void. (b) Schematic of variations in pore shapes: void aspect ratio from 0.01 to 1.

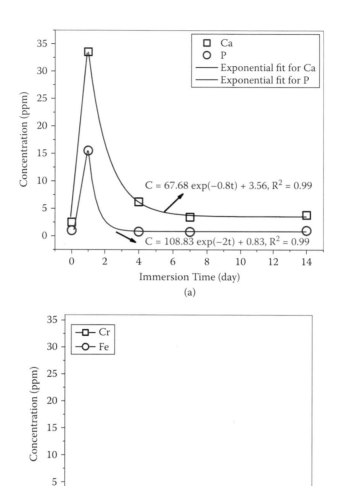

$$C = 67.68 \exp(-0.8t) + 3.56, R^2 = 0.99$$

$$C = 108.83 \exp(-2t) + 0.83, R^2 = 0.99$$

(a)

(b)

COLOR FIGURE 5.3
(a) Change of Ca and P concentration with time of immersion shows the dissolution followed by a deposition of CaP compounds. (C and t stand for concentration and immersion time, respectively.) (Reprinted from Dey and Mukhopadhyay, *International Journal of Applied Ceramic Technology*, 11: 65–82, 2013. With permission from American Ceramic Society and Wiley.) (b) Chromium and iron leaching show almost nil concentration even after 14 days of accelerated dissolution test in a SBF environment.

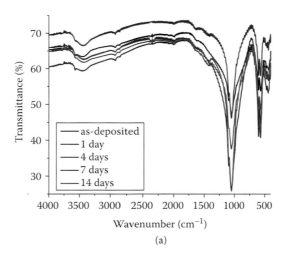

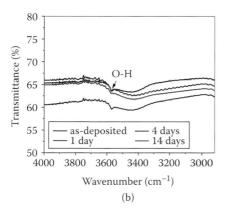

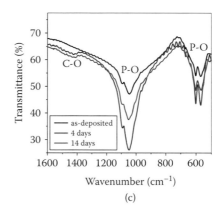

(a)

(b)   (c)

COLOR FIGURE 5.5
(a) FTIR spectra of the MIPS-HAp coating before and after SBF immersion. Enlarged view showing the location of (b) OH and (b) CO₃ and PO₄ characteristic peaks. (Reprinted from Dey and Mukhopadhyay, *International Journal of Applied Ceramic Technology*, 11: 65–82, 2013. With permission from American Ceramic Society and Wiley.)

COLOR FIGURE 6.1
A typical image of the MIPS-HAp coating on SS316L cylindrical substrate (both diameter and length of ~25 mm) of different thicknesses, e.g., 100, 200, and 300 μm (on left side onward), and grit-blasted uncoated SS316L cylindrical substrate before the coating (on extreme right side image).

COLOR FIGURE 6.2
Arrangements before bonding strength measurement: two identical SS316L cylindrical stubs (e.g., top one coated and bottom one uncoated) joined with adhesive tape and cured in oven for several hours.

COLOR FIGURE 6.4
Arrangement for three-point bending test of MIPS-HAp coating on SS316L substrate (HAp coated surface is kept in the bottom, i.e., tensile side).

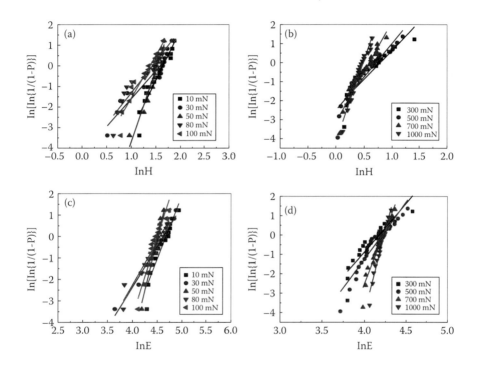

COLOR FIGURE 7.3
Weibull distribution plots of nanohardness at (a) low (10–100 mN) and (b) high (300–1000 mN) loads. Weibull distribution plots of Young's modulus at (c) low (10–100 mN) and (d) high (300–1000 mN) loads.

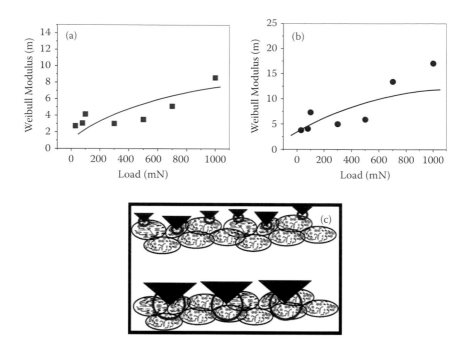

COLOR FIGURE 7.4
Weibull modulus of (a) nanohardness and (b) Young's modulus as a function of load of the MIPS-HAp coatings. (c) Model of indenter-MIPS-HAp coating interaction. (Reprinted/modified from Dey et al., *Ceramics International*, 35: 2295–2304, 2009. With permission from Elsevier. Reprinted/modified from Dey et al., *Journal of Materials Science*, 44: 4911–4918, 2009. With permission from Springer.)

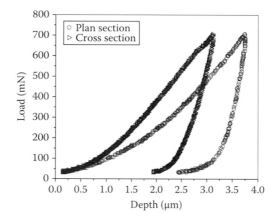

COLOR FIGURE 7.7
(a) Load-depth plots from nanoindentation experiments on MIPS-HAp coating at 700 mN on plan section and cross section. Ratio of (b) nanohardness data of the cross section (H_{cs}) and plan section (H_{ps}) as a function of load (replotted from Dey and Mukhopadhyay, *Advances in Applied Ceramics*, 109: 346–354, 2010) and (c) Young's modulus data of the cross section (E_{cs}) and Young's modulus data of the plan section (E_{ps}) as a function of load.

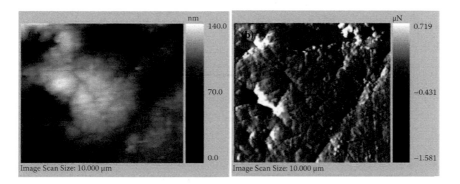

COLOR FIGURE 8.2
Coefficient of friction of the MIPS-HAp coating as a function of scratch time. Red line represents the average of five scratch test data.

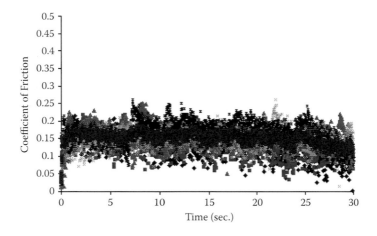

COLOR FIGURE 8.3
In situ SPM photomicrographs (image scan size: 10×10 μm) of the region of the MIPS-HAp coating on which the nanoscratch experiments were performed (a) in topographical and (b) in gradient mode, showing that no peel-off occurred after the 100 μN constant normal force nanoscratch experiments.

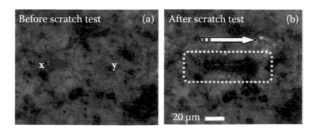

COLOR FIGURE 8.5
Optical micrographs of the MIPS-HAp coating (a) before (starting and end points of the scratch test marked x and y, respectively) and (b) after (the arrow indicates the direction of scratching) the nanoscratch testing by ramping the normal force from 0 to 700 mN. (Reprinted/modified from Dey et al., *Journal of Thermal Spray Technology*, 18: 578–592, 2009. With permission from Springer.)

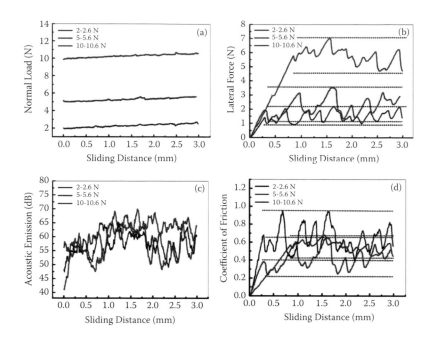

COLOR FIGURE 8.6
Data on the variations of (a) normal ramping load, (b) lateral force, (c) acoustic emission, and (d) coefficient of friction as a function of sliding distance for the as-sprayed MIPS-HAp coating.

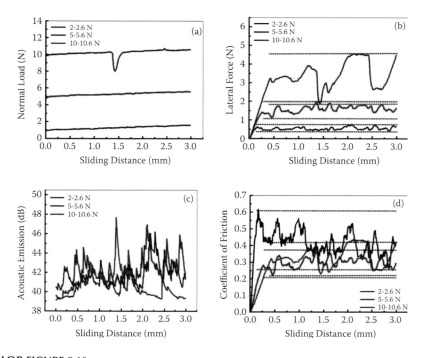

COLOR FIGURE 8.10
Data on the variations of (a) normal ramping load, (b) lateral force, (c) acoustic emission, and (d) coefficient of friction as a function of sliding distance for the polished MIPS-HAp coating.

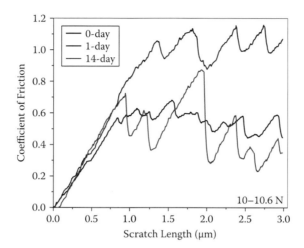

COLOR FIGURE 8.13
Variations of the coefficient of friction (COF) of the MIPS-HAp coating as a function of sliding distance before and after immersion in the SBF solution. (Reprinted/modified from Dey and Mukhopadhyay, *International Journal of Applied Ceramic Technology*, 11: 65–82, 2013. With permission from Wiley.)

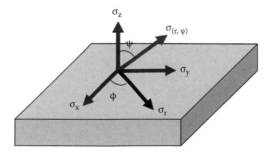

COLOR FIGURE 9.1
Schematic representation of stresses acting on a residually stressed body. (Adopted/modified and redrawn from Yang et al., *Biomaterials*, 21: 1327–1337, 2000. With permission from Elsevier.)

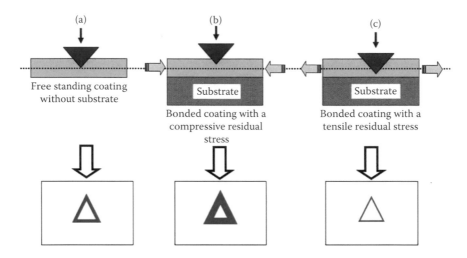

COLOR FIGURE 9.2
Schematic of nanoindentation processes on a (a) freestanding coating without substrate and bonded coating with a (b) compressive and (c) tensile stress.

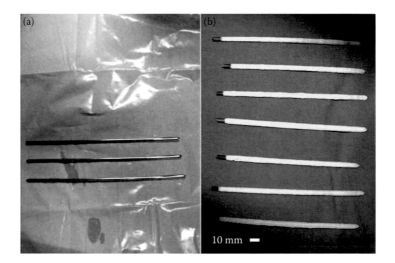

10 mm ▬

COLOR FIGURE 10.1
SS316L intramedullary pins: (a) uncoated and (b) MIPS-HAp coated.

COLOR FIGURE 10.2
Progressive steps (marked as arrows from right to left) of implantation of MIPS-HAp-coated intramedullary pin into the New Zealand white rabbit: deliberate creation of defect by micro-drilling, closing muscle layers, and closure out of the skin.

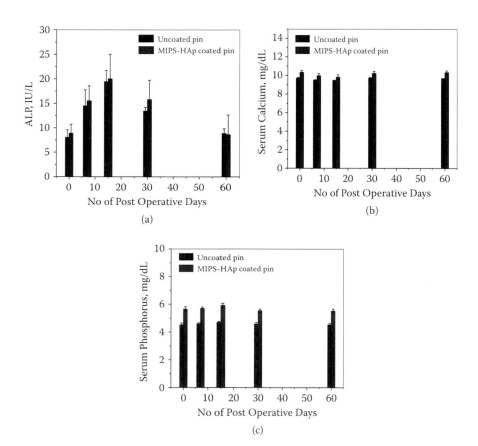

COLOR FIGURE 10.3
Changes in (a) serum alkaline phosphatase activity (IU/L), serum calcium estimation (mg/dl), and serum phosphorous level (mg/dl) at different durations of fracture healing for animals inserted with uncoated and MIPS-HAp-coated pins. (Reprinted/modified from Dey et al., *Ceramics International*, 8: 1377–1391, 2011. With permission from Elsevier.)

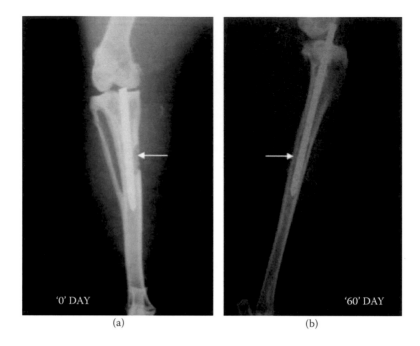

(a)   (b)

COLOR FIGURE 10.4
Radiographic observation at the fracture site, taken at (a) 0 and (b) 60 days post-operatively for the MIPS-HAp-coated pins. The arrow in (a) indicates the artificial defects created in bone, and the arrow in (b) indicates the complete annihilation of the bone defect by newly formed osseous tissue, and thus established cortical continuity. (Reprinted/modified from Dey et al., *Ceramics International*, 8: 1377–1391, 2011. With permission from Elsevier.)

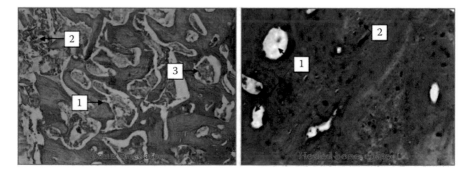

COLOR FIGURE 10.5
Histopathology photomicrographs of (a) uncoated and (b) MIPS-HAp-coated pins inserted in rabbits. (Reprinted/modified from Dey et al., *Ceramics International*, 8: 1377–1391, 2011. With permission from Elsevier.)

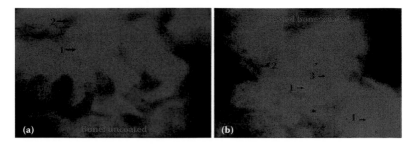

COLOR FIGURE 10.6
Fluorochrome labeling study: photomicrograph of (a) uncoated and (b) MIPS-HAp-coated pins inserted in rabbits. (Reprinted/modified from Dey et al., *Ceramics International*, 8: 1377–1391, 2011. With permission from Elsevier.)

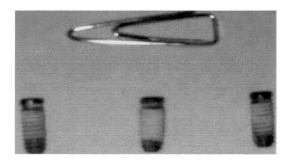

COLOR FIGURE 11.1
MIPS-HAp-coated Ti-6Al-4V dental implants (screw).

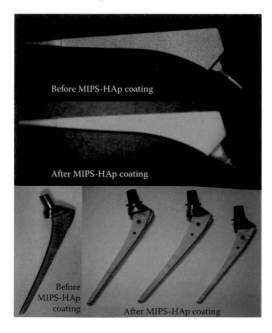

COLOR FIGURE 11.2
Before (grit blasted) and after MIPS-HAp coating on different-shaped SS316L hip implants (length, 155 mm for largest size, 140 mm for smallest size).

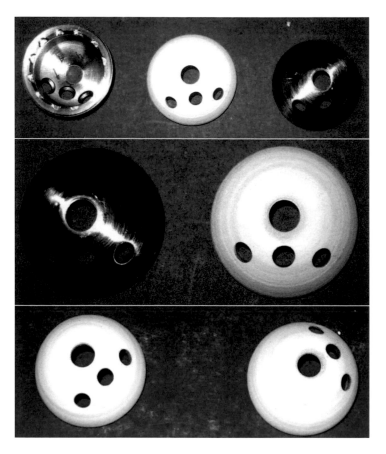

COLOR FIGURE 11.3
Before and after MIPS-HAp coating on SS316L acetabular shell (outer diameter, 48.35; inner diameter, 39.75 mm).

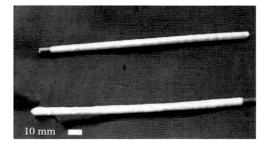

COLOR FIGURE 11.4
MIPS-TCP-coated intramedullary pin. (Reprinted/modified from Dey et al., *Ceramics International*, 8: 1377–1391, 2011. With permission from Elsevier.)

What happens in a sol-gel process is that the suspended particles in a sol agglomerate, and under controlled conditions, they eventually link together to form a gel. In the perspective of material synthesis as such, there are two generic variations of the sol-gel technique. One is called the colloidal method, as mentioned above. The other is called the polymeric (or alkoxide) route. It uses an alcohol to suspend the desired particles in the sol. The activation of the precursor can be conveniently achieved by the addition of either an acid, e.g., HCl, or a base, e.g., KOH. The activated precursors then react together to form a network. The network grows and ages with time and temperature until it is the size of the container. At this point, the viscosity of the liquid increases at an exponential rate until gelation occurs; that is, no more flow is observed. As mentioned earlier, the most widespread commercial use of sol-gel is in the fabrication and deposition of coatings. The role of the coating may be to provide one or more of a number of functions. These may include, for instance, but are not necessarily limited to, corrosion protection, abrasion resistance, antireflective performance, smart windows, provision of super hydrophobic/hydrophilic surface, ferroelectric activity, surface conductivity, and catalytic activity through very high surface area generation, etc. Nevertheless, the major limitation of this process is that usually only very thin films (few nanometers/angstroms thick) with appropriate quality can be produced by it, and attempts to produce thicker films usually fail due to obvious cracking that results from the differential shrinkage rate during the drying stage.

2.7.9 Biomimetic Coating Process

What is biomimetics? Well, it literally means the mimicry of biology. It is a branch of science in which biologists and engineers shake hands to produce "bioinspired" materials. Recent research [82–84] has shown that the biomimetic process is also one of the most promising techniques for producing a bioactive coating at ambient temperature. This method aids in natural-like formation of a biologically active apatite layer. The layer is usually formed on a substrate, e.g., a biomedical implant surface. However, the formation of the active apaptite layer happens after immersion of the substrate into an artificially prepared solution of calcium and phosphate. The solution is usually kept supersaturated at a temperature of about 37°C and a pH of about 7.4. This supersaturated solution has a composition identical to that of the fluid inside a human body. Hence, this solution is known as simulated body fluid (SBF). What happens in this biomimetic deposition process is that through a heterogeneous nucleation process, a classical biomimetic calcium phosphate phase grows out of the SBF solution and nucleates and grows as a coating on the immersed substrate within about 2 to 4 weeks of immersion, provided the SBF solution is appropriately replenished. Although the phase purity can be assured from this process, due to its very low deposition rate, it is not viable for commercialization.

2.8 Microplasma vs. Macroplasma Spraying

Truly speaking, the microplasma spraying (MIPS) process is basically an engineered modification of the conventional macroplasma spraying (MAPS) process. Already, it has been discussed that as the plasma spot diameter is within 1 to 5 mm, the MIPS process can do very precision-level coating that the high-power plasma gun used in the MAPS process cannot do. Thus, in the MIPS process, the components or implants like dental implants, intramedullary pins for bone fracture fixation, etc., with very small dimensions can be sprayed with high spray efficiency while still maintaining the desired thickness. Considering process parameters of MIPS, it is possible to feed finer powder particle. In the MIPS process, the plasmatron power (e.g., 1–4 kW) is characteristically and deliberately kept much lower than that (e.g., 10–40 kW) of the MAPS process. Since less power input is involved, the chance of the powder overheating, and hence of an undesirable phase change of the HAp powder, as well as overheating of the underlying metallic substrate being coated, is maximally minimized in the MIPS process. Further, it will also contribute to minimize the magnitude of thermally induced stress. Because of the efficient miniaturization, the dimensions of the MIPS facility can fit in to a "tabletop" facility, and yet can be tailored to still remain portable to different zones of the workplace [42–57]. As the MIPS facility is compact and portable, it can be operated in a clean-room environment with trivial modifications. Moreover, the noise level (25–50 dB) of the MIPS equipment is much lower than that (80–120 dB) [25–27] of the MAPS equipment. Very recent work by the present authors' research group has shown that the MIPS process can successfully produce highly crystalline yet adequately porous phase-pure HAp coatings on biomedical implants [30–41].

2.9 Summary

In the present chapter, we have discussed the basics of the plasma spraying process. It has been illustrated how a coating formation happens by this process. The relative advantages of the MIPS process over the MAPS process have been adequately addressed. The uniqueness of the MIPS process as a coating manufacturing technique has been highlighted, along with its application. The other coating processes were also discussed, along with a presentation of a comparative narration of their relative merits and demerits.

HAp coating and its applications will be discussed in Chapter 3. The influence of plasma spraying parameters on HAp coating's characteristics will be elaborated. Further, the relative merits and demerits of high-temperature- and low-temperature-assisted HAp coating deposition processes will be critically analyzed. The state-of-the-art scenario of the nanostructured HAp coatings, HAp composite coatings, and doped HAp coatings will be described with an analysis of their relative limitations and advantages. Of course, a major emphasis will be given on the relative efficacies of the HAp coatings deposited by the MAPS and MIPS processes. This discussion will finally highlight the major success of MIPS-HAp coatings developed in recent times.

References

1. Gross K. A. and Berndt C. C. 2002. Biomedical application of apatites. *Reviews in Mineralogy and Geochemistry*, 48: 631–672.
2. Hench L. L. and Ethridge E. C. 1982. *Biomaterials: an interfacial approach*. New York: Academic Press.
3. Hench L. L. 1993. Bioceramics: from concept to clinic. *American Ceramic Society Bulletin*, 72: 93–98.
4. Khudonogov G. I. 1994. Microplasma technology for applying aesthetic coatings to fixed dental prostheses. *Meditsinskaya Tekhnika*, 6: 37–38.
5. Yushchenko K., Borisov Y., Pereverzev Y., Vojnarovitch S., Darmochval V., Bobric V., Ramaekers P., and Raa G. 1998. Microplasma spraying. *Proceedings of the International Thermal Spray Conference*, 2: 1461–1467.
6. Bogachek V. L. and Tsypina L. N. 1990. Quality of coatings produced by microplasma spraying. *Welding International*, 4: 14–16.
7. Bo S., Yongfeng B., and Changjiu L. 1999. Effect of spray parameters on the microstructure and properties of microplasma sprayed NiCrBSi coatings. *International Conference on Advanced Manufacturing Technology '99*, 1287–1290.
8. Gao Y., Xu X., Yan Z., and Xin G. 2002. High hardness alumina coatings prepared by low power plasma spraying. *Surface and Coatings Technology*, 154: 189–193.
9. Li C. and Sun B. 2003. Microstructure and property of Al_2O_3 coating microplasma-sprayed using a novel hollow cathode torch. *Materials Letters*, 58: 179– 183.
10. Li C. and Sun B. 2004. Microstructure and property of micro-plasma-sprayed Cu coating. *Materials Science and Engineering A*, 379: 92–101.
11. Li C.-J., Sun B., Wang M., Han F., and Wu T. 2003. Microstructure and properties of micro-plasma sprayed Cu coating. In *Thermal spray 2003: advancing the science and applying the technology*, Moreau C. and Marple B. (Eds.). Materials Park, OH: ASM International.
12. Crawford W. S., Cappelli M. A., and Prinz F. B. 2005. Design and tuning of a vacuum microplasma spray system: particle entrainment. In *Surface engineering in materials science III*, Agarwal A., Dahotre N. B., Seal S., Moore J. J., and Blue C. (Eds.). Minerals, Metals and Materials Society, 267–282.

13. Zhao L., Bobzin K., Ernst F., Zwick J., and Lugscheider E. 2006. Study on the influence of plasma spray processes and spray parameters on the structure and crystallinity of hydroxylapatite coatings. *Materialwissenschaft und Werkstofftechnik*, 37: 516–520.

14. Lugscheider E., Bobzin K., Zhao L., and Zwick J. 2006. Assessment of the microplasma spraying process for the coating application. *Advanced Engineering Materials*, 8: 635–639.

15. Shimizu Y., Sasaki T., Bose A. C., Terashima K., and Koshizaki N. 2006. Development of wire spraying for direct micro-patterning via an atmospheric-pressure UHF inductively coupled microplasma jet. *Surface and Coatings Technology*, 200: 4251–4256.

16. Shimizu Y., Sasaki T., Terashima K., and Koshizaki N. 2006. Localized deposition technique using an atmospheric-pressure microplasma jet for on-demand material processing. *Journal of Photopolymer Science and Technology*, 19: 235–240.

17. Bose A. C., Shimizu Y., Mariotti D., Sasaki T., Terashima K., and Koshizaki N. 2006. Flow rate effect on the structure and morphology of molybdenum oxide nanoparticles deposited by atmospheric-pressure microplasma processing. *Nanotechnology*, 17: 5976–5982.

18. He D., Zhao Q., Zhao L., and Sun X. 2007. Influence of microplasma spray parameters on the microstructure and crystallinity of hydroxyapatite coatings. *Chinese Journal of Materials Research*, 2007, 6, 659–663.

19. Hua S.-C., Wang H.-G., Wang L.-Y., and Liu G. 2007. Optimization of the process parameters of nanostructured AT13 coatings prepared by micro-plasma spraying. *Journal of Inorganic Materials*, 22: 560–564.

20. Liu G., Wang L. Y., Wang H. G., and Hua S. C. 2007. Properties of copper coatings deposited by multi-function microplasma spraying. In *Thermal spray 2007: global coating solutions: proceedings of the 2007*, Marple B. R., Hyland M. M., Lau Y.-C., Li C.-J., Lima R. S., and Montavon G. (Eds.). Materials Park, OH: ASM International, 1007–1010.

21. Lei M. K., Zhu X. P., Xu K. W., and Xu B. S. 2008. Properties of multi-function micro-plasma sprayed nanostructured Al_2O_3-13wt%TiO_2 coatings. *Key Engineering Materials*, 373–374: 59–63.

22. Hua S. C., Wang H. G., Wang L. Y., and Liu G. 2008. Microstructure and properties of multi-function micro-plasma sprayed Al_2O_3 coatings. *Key Engineering Materials*, 373–374: 51–54.

23. Liu G., Wang L.-Y., Chen G.-M., Wei W.-N., Hua S.-C., and Zhu E.-L. 2001. Preparation and properties of SiC-CNTs/Al_2O_3-TiO_2 coating. *Journal of Inorganic Materials*, 26: 1187–1192.

24. Zhao Q., He D., Zhao L., and Li X. 2011. In-vitro study of microplasma sprayed hydroxyapatite coatings in Hanks balanced salt solution. *Materials and Manufacturing Processes*, 26: 175–180.

25. Junker R., Manders P. J. D., Wolke J., Borisov Y., and Jansen J. A. 2010. Bone-supportive behavior of microplasma-sprayed CaP-coated implants: mechanical and histological outcome in the goat. *Clinical Oral Implants Research*, 21: 189–200.

26. Junker R., Manders P. J. D., Wolke J., Borisov Y., and Jansen J. A. 2010. Bone reaction adjacent to microplasma sprayed CaP-coated oral implants subjected to occlusal load, an experimental study in the dog. Part I. Short-term results. *Clinical Oral Implants Research*, 21: 1251–1263.

27. Junker R., Manders P. J. D., Wolke J., Borisov Y., and Jansen J. A. 2011. Bone reaction adjacent to microplasma sprayed calcium phosphate-coated oral implants subjected to an occlusal load, an experimental study in the dog. *Clinical Oral Implants Research*, 22: 135–142.
28. Liu G., Wang L., Chen G., Wei W., Hua S., and Zhu E. 2011. Effect of spraying parameters on the microstructure and mechanical properties of micro-plasma sprayed alumina-titania coatings. *Plasma Science and Technology*, 13: 474–479.
29. Mistry S., Kundu D., Datta S., and Basu D. 2011. Comparison of bioactive glass coated and hydroxyapatite coated titanium dental implants in the human jaw bone. *Australian Dental Journal*, 56: 68–75.
30. Dey A., Mukhopadhyay A. K., Sinha, Gangadharan S., Sinha M. K., Basu D., and Bandyopadhyay N. R. 2009. Nanoindentation study of microplasma sprayed hydroxyapatite coating. *Ceramics International*, 35: 2295–2304.
31. Dey A., Mukhopadhyay A. K., Gangadharan S., Sinha M. K., and Basu D. 2009. Characterization of microplasma sprayed hydroxyapatite coating. *Journal of Thermal Spray Technology*, 18: 578–592.
32. Dey A., Mukhopadhyay A. K., Gangadharan S., Sinha M. K., and Basu D. 2009. Development of hydroxyapatite coating by microplasma spraying. *Materials and Manufacturing Processes*, 24: 1321–1330.
33. Dey A., Mukhopadhyay A. K., Sinha, Gangadharan S., Sinha M. K., and Basu D. 2009. Weibull modulus of nano-hardness and elastic modulus of hydroxyapatite coating. *Journal of Materials Science*, 44: 4911–4918.
34. Dey A., and Mukhopadhyay A. K. 2010. Anisotropy in nano-hardness of microplasma sprayed hydroxyapatite coating. *Advances in Applied Ceramics*, 109: 346–354.
35. Dey A., Mukhopadhyay A. K., Sinha M. K., and Basu D. 2010. Characterization of plasma sprayed porous hydroxyapatite coating. *Journal of the Institution of Engineers (India), Part MM, Metallurgy and Material Science Division*, 91: 3–7.
36. Dey A., Nandi S. K., Kundu B., Kumar C., Mukherjee P., Roy S., Mukhopadhyay A. K., Sinha M. K., and Basu D. 2011. Evaluation of hydroxyapatite and β-tri calcium phosphate microplasma spray coated pin intra-medullarly for bone repair in a rabbit model. *Ceramics International*, 8: 1377–1391.
37. Dey A., and Mukhopadhyay A. K. 2011. Fracture toughness of microplasma sprayed hydroxyapatite coating by nanoindentation. *International Journal of Applied Ceramic Technology*, 8: 572–590.
38. Dey A. and Mukhopadhyay A. K. 2014. Evaluation of residual stress in microplasma sprayed hydroxyapatite coating by nanoindentation. *Ceramics International*, 40A: 1263–1272.
39. Dey A. and Mukhopadhyay A. K. 2013. In-vitro dissolution, microstructural and mechanical characterizations of microplasma sprayed hydroxyapatite coating. *International Journal of Applied Ceramic Technology*, 11: 65–82.
40. Dey A., Banerjee K., and Mukhopadhyay A. K. 2014. Microplasma sprayed hydroxyapatite coating: emerging technology for biomedical application. *Materials Technology*, 29: B10–B15.
41. Dey A., Sinha A., Banerjee K., and Mukhopadhyay A. K. 2014. Tribological studies of microplasma sprayed hydroxyapatite coating at low load. *Materials Technology*, 29: B35–B40.

42. Zajchowski P. H., Blankenship D. R., and Shubert G. C. 2006. Methods for repairing workpieces using microplasma spray coating. EP1652952 A1.
43. Zajchowski P. H., Blankenship D. R., and Shubert G. C. 2006. Method and apparatus for repairing thermal barrier coatings. EP1652955 A1.
44. Blankenship D. R., Shubert G. C., and Zajchowski P. H. 2006. Microplasma spray coating apparatus. US7115832 B1.
45. Blankenship D. R., Zajchowski P. H., and Shubert G. C. 2006. Plasma spray apparatus. US20060091117 A1.
46. Zajchowski P. H., Blankenship D. R., and Shubert G. C. 2006. Method and apparatus for repairing thermal barrier coatings. US20060091119 A1.
47. Zajchowski P. H., Blankenship D. R., and Shubert G. C. 2007. Methods for repairing workpieces using microplasma spray coating. US20070023402 A1.
48. Blankenship D. R., Rutz D. A., Pietrusca N. A., Zajchowski P. H., and Shubert G. C. 2007. Microplasma deposition apparatus and methods. EP1743951 A2.
49. Blankenship D. R., Shubert G. C., and Zajchowski P. H. 2008. Microplasma spray coating apparatus. EP1749583 B1.
50. Blankenship D. R., Memmen R. L., Shubert G. C., and Zajchowski P. H. 2009. Method for microplasma spray coating a portion of a turbine vane in a gas turbine engine, EP1652951 B1.
51. Blankenship D. R., Shubert G. C., and Zajchowski P. H. 2009. Method and apparatus for microplasma spray coating a portion of a compressor blade in a gas turbine engine. EP1652954 B1.
52. Blankenship D. R., Zajchowski P. H., and Shubert G. C. 2009. Methods for repairing a workpiece. US20090208662 A1.
53. Zajchowski P. H., Blankenship D. R., Shubert G. C., and Memmen R. L.2009. Method and apparatus for microplasma spray coating a portion of a turbine vane in a gas turbine engine. US20090314202 A1.
54. Zajchowski P. H., Blankenship D. R., and Shubert G. C. 2010. Method and apparatus for microplasma spray coating a portion of a compressor blade in a gas turbine engine. US20100199494 A1.
55. Blankenship D. R., Zajchowski P. H., and Shubert G. C. 2010. Microplasma spray apparatus and method for coating articles using same. US20100200549 A1.
56. Blankenship D. R., Rutz D. A., Pietruska N. A., Zajchowski P. H., and Shubert. 2011. Deposition apparatus and methods. US8067711 B2.
57. Lee G., Yadav O. P., and Lim D. 2011. Repair of a coating on a turbine component. US20110206533 A1.
58. Khor K. A., Li H., and Cheang P. 2003. Characterization of the bone-like apatite precipitated on high velocity oxy-fuel (HVOF) sprayed calcium phosphate deposits. *Biomaterials*, 24: 769–775.
59. Li H., Khor K. A., and Cheang P. 2002. Young's modulus and fracture toughness determination of high velocity oxy-fuel-sprayed bioceramic coatings. *Surface and Coatings Technology*, 155: 21–32.
60. Li H., Khor K. A., and Cheang P. 2002. Titanium dioxide reinforced hydroxyapatite coatings deposited by high velocity oxy-fuel (HVOF) spray. *Biomaterials*, 23: 85–91.

61. Mohammadi Z. A., Moayyed A. Z., and Mesgar A. S. M. 2007. Adhesive and cohesive properties by indentation method of plasma-sprayed hydroxyapatite coatings. *Applied Surface Science*, 253: 4960–4965.
62. Gu Y. W., Khor K. A., and Cheang P. 2003. In vitro studies of plasma-sprayed hydroxyapatite/Ti-6Al-4V composite coatings in simulated body fluid (SBF). *Biomaterials*, 24: 1603–1611.
63. Garcia-Sanz F. J., Mayor M. B., Arias J. L., Pou J., Leon B., and Perez-Amor M. 1997. Hydroxyapatite coatings: a comparative study between plasma-spray and pulsed laser deposition techniques. *Journal of Materials Science: Materials in Medicine*, 8: 861–865.
64. Cheng G. J., Pirzada D., Cai M., Mohanty P., and Bandyopadhyay A. 2005. Bioceramic coating of hydroxyapatite on titanium substrate with Nd-YAG laser. *Materials Science and Engineering C*, 25: 541–547.
65. Chen Y. Y., Zhang Q., Zhang T. H., Gan C. H., Zheng C. Y., and Yu G. 2006. Carbon nanotube reinforced hydroxyapatite composite coatings produced through laser surface alloying. *Carbon*, 44: 37–45.
66. Arias J. L., Mayor M. B., Pou J., Leng Y., Leon B., and Perez-Amora M. 2003. Micro- and nano-testing of calcium phosphate coatings produced by pulsed laser deposition. *Biomaterials*, 24: 3403–3408.
67. Abdullah C., Noorakma W., Zuhailawati H., Aishvarya V., and Dhindaw B. K. 2013. Hydroxyapatite-coated magnesium-based biodegradable alloy: cold spray deposition and simulated body fluid studies. *Journal of Materials Engineering and Performance*, 22: 2997–3004
68. Lu L., Zhou X., and Mohanty P. 2014. In vitro immersion behavior of cold sprayed hydroxyapatite/titanium composite coatings. *Journal of Biosciences and Medicines*, 2: 10–16.
69. Huang Y., Song L., Liu X., Xiao Y., Wu Y., Chen J., Wu F., and Gu Z. 2010. Hydroxyapatite coatings deposited by liquid precursor plasma spraying: controlled dense and porous microstructures and osteoblastic cell responses. *Biofabrication*, 2: 045003.
70. Huang Y., Song L., Liu X., Xiao Y., Wu Y., Chen J., Wu F., and Gu Z. 2010. Characterization and formation mechanism of nano-structured hydroxyapatite coatings deposited by the liquid precursor plasma spraying process. *Biomedical Materials*, 5: 054113.
71. Guo X., Gough J., and Xiao P. 2007. Electrophoretic deposition of hydroxyapatite coating on Fecralloy and analysis of human osteoblastic cellular response. *Journal of Biomedical Materials Research A*, 80: 24–33.
72. Ma J., Wang C., and Peng K. W. 2003. Electrophoretic deposition of porous hydroxyapatite scaffold. *Biomaterials*, 24: 3505–3510.
73. Li X., Huang J., and Edirisinghe M. 2008. Development of nano-hydroxyapatite coating by electrohydrodynamic atomization spraying. *Journal of Materials Science: Materials in Medicine*, 19: 1545–1551.
74. Huang J. S., Jayasinghe N. S., Best M., Edirisinghe M. J., Brooks R. A., and Bonfield W. 2004. Electrospraying of a nano-hydroxyapatite suspension. *Journal of Materials Science*, 39: 1029–1032.

75. Cheng K., Zhang S., Weng W., Khor K. A., Miao S., and Wang Y. 2008. The adhesion strength and residual stress of colloidal-sol gel derived [beta]-tricalcium-phosphate/fluoridated-hydroxyapatite biphasic coatings. *Thin Solid Films*, 516: 3251–3255.

76. Liu D. M., Troczynski T., and Tseng W. J. 2001. Water-based sol-gel synthesis of hydroxyapatite: process development. *Biomaterials*, 22: 1721–1730.

77. Kim H. W., Yoon B. H., Koh Y. H., and Kim H. E. 2006. Processing and performance of hydroxyapatite/fluorapatite double layer coating on zirconia by the powder slurry method. *Journal of the American Ceramic Society*, 89: 2466–2472.

78. Kim H. W., Georgiou G., Knowles J. C., Koh Y. H., and Kim H. E. 2004. Calcium phosphates and glass composite coatings on zirconia for enhanced biocompatibility. *Biomaterials*, 25: 4203–4213.

79. Manso-Silvan M., Langlet M., Jimenez C., Fernandez M., and Martinez-Duart J. M. 2003. Calcium phosphate coatings prepared by aerosol-gel. *Journal of the European Ceramic Society*, 23: 243–246.

80. Weng W. and Baptista J. L. 1999. Preparation and characterization of hydroxyapatite coatings on Ti6Al4V alloy by a sol-gel method. *Journal of the American Ceramic Society*, 82: 27–32.

81. Balamurugan A., Balossier G., Kannan S., Michel J., Faure J., and Rajeswari S. 2007. Electrochemical and structural characterisation of zirconia reinforced hydroxyapatite bioceramic sol-gel coatings on surgical grade 316L SS for biomedical applications. *Ceramics International*, 33: 605–614.

82. Kim H. M., Miyaji F., Kokubo T., and Nakamura T. 1997. Bonding strength of bonelike apatite layer to Ti metal substrate. *Journal of Biomedical Materials Research Part B*, 38: 121–127.

83. Habibovic P., Barrere F. C., vanBlitterswijzk A., de Groot K., and Layrolle P. 2002. Biomimetic hydroxyapatite coating on metal implants. *Journal of the American Ceramic Society*, 85: 517–522.

84. Chakraborty J., Sinha M. K., and Basu D. 2007. Biomolecular template induced biomimetic coating of hydroxyapatite on an SS316L substrate. *Journal of the American Ceramic Society*, 90: 1258–1261.

85. Nieh T. G., Jankowsk A. F., and Koike J. 2001. Processing and characterization of hydroxyapatite coatings on titanium produced by magnetron sputtering. *Journal of Materials Research*, 16: 3238–3245.

86. Nieh T. G., Choi B. W., and Jankowski A. F. 2001. Synthesis and characterization of porous hydroxyapatite and hydroxyapatite coatings. Report submitted to Minerals, Metals and Materials Society Annual Meeting and Exhibition, Los Angeles.

87. Li X., Huang J., and Edirisinghe M. J. 2008. Novel patterning of nano-bioceramics: template assisted electrohydrodynamic atomisation spraying. *Journal of Royal Society Interface*, 5: 253–257.

3

Hydroxyapatite Coating and Its Application

In the present chapter, we discussed the hydroxyapatite (HAp) coating and its applications. You may ask why HAp coating is required at all. You may also inquire about the factors that influence the properties of HAp coatings and so on. We shall attempt in this chapter to provide some answers to such queries.

3.1 Background of the Problem and Basic Issues

Now, we want to tell a story that explains why HAp coatings become essential in orthopedic applications in particularly bone fracture fixations. The problem originated from a critical biomedical engineering issue, e.g., loosening of metallic prosthesis fixed with the polymethylmethacrylate bone cement, especially in the case of hip joint replacement, which ultimately causes the patient to undergo painstaking revision surgery. Then the engineers produced a cementless fixation introducing a bioactive HAp (i.e., the main inorganic constituent of bones and teeth) coating on the metallic implants. A wide variety of different coating methods have been developed to make the HAp coating on metallic implants more reliable, of which ultimately the plasma spraying method has been commercially accepted. However, the story was not yet finished at all; a lot of questions came out regarding coating adherence, stability, and biofunctionality in both in vitro and in vivo environments. Moreover, it has been now realized that the conventional high-power plasma spray (i.e., macroplasma) coating method renders plenty of disadvantages. The major limitations are poor crystallinity, phase impurity, inadequate porosity that hinders osseointegration, and residual stresses, which ultimately lead to inadequate mechanical properties and delamination of the coating. These limitations actually lead to the very recent development of the microplasma spraying method, as mentioned in Chapter 2.

3.2 Applications of HAp Coating

Bone is a natural ceramic composite of HAp-reinforced collagen. This is why in the biomedical field the bioactive ceramic coatings, e.g., HAp coatings, are used on metallic implants in:

1. Total hip joint prostheses as an alternative to PMMA cement-based fixation
2. Dental implants for bioactive fixation
3. Fillers for repairing bone defects

In all these applications, the highly convoluted interface that develops can offer excellent mechanical stability for the bone to grow into the pores of the ceramic because it accelerates the rate of bone growth and enhances the strength of the bone-implant interface. Thus, the bioactive HAp coating accelerates the kinetics of osseointegration, shields the orthopedic implant from environmental attack, e.g., corrosion, and stops the fibrous tissue ingrowth that loosens the implant from the damaged tissue site.

3.3 HAp Coating Developed by Different Methods

As already mentioned in Chapter 2, there are many methods reported for preparation of HAp or HAp composite coatings, including their in vitro and in vivo testing [1–37]. These include:

1. Conventional macroplasma spraying (MAPS) [1–7]
2. Microplasma spraying (MIPS) [8–18]
3. Laser-assisted processes [19–24]
4. Other thermal spraying techniques [25–27]
5. Sputtering [28, 29]
6. Electrophoretic deposition (EPD) techniques [30, 31]
7. Sol-gel (SG)-derived techniques [32, 33]
8. Biomimetics [34, 35]
9. Electrohydrodynamic atomization (EHDA) spraying [36, 37]
10. Dip/slurry coatings [38], etc.

3.3.1 Issues Related to High-Temperature Processes

There are mainly two types of high-temperature coating processes: thermal spraying techniques and laser-assisted processes. Both of these processes suffer from the limitation that in addition to the desirable HAp phase, undesirable amorphous as well as other crystalline phases, such as calcium oxide (CaO), tricalcium phosphate (TCP), and tetracalcium phosphate (TTCP), are formed as a result of exposure to high temperatures. The major problem is that when the coating is exposed to physiological fluids, these additional phases prove to be detrimental for the long-term chemical stability of the coating [7]. The other major limitation of any high-temperature process is that the higher the temperature, the greater the chance to have larger magnitudes of frozen-in residual stresses. The higher the residual stress magnitude, the larger the probability of coating failure due to delamination at a later stage [39–42]. That is why, notwithstanding its drawbacks, the plasma spraying process is still the most widely used method for commercial purposes. The major reasons for the same are high deposition rate, better coating adhesion, and possibility to control coating thickness.

3.3.2 Issues Related to Low-Temperature Processes

As already discussed in Chapter 2, the major low-temperature coating processes are biomimetics, sol-gel (SG) derived, electrohydrodynamic atomization (EHDA) spraying, and dip/slurry coatings. These techniques produce uniform coatings not only on any type of substrate material (e.g., metals, ceramics, and polymers), but also on even porous or complex implant geometries. It has been further demonstrated that high-quality novel nano-hydroxyapatite (nHAp) coatings can be produced by biomimetics [35] and EHDA [36, 37] techniques. But all these processes, including the EPD technique, have two major limitations. The first limitation is that coating adhesion is far from desirable quality. The second limitation is that it is very difficult to tune and control the porosity to the desirable level in coatings produced by such techniques. The coatings produced by SG-derived and EHDA techniques are unacceptably thin. Further, it takes an almost unacceptably long time, e.g., 2 to 4 weeks, to deposit a biomimetic coating [34, 35]. There is also a major problem of cracking due to densification in coatings deposited by the EPD technique [30]. Cracking during the drying stage is also a major limitation of the coatings prepared by the SG-derived techniques [32, 33]. Except for the biomimetics and EHDA techniques, all other coating deposition techniques require post-heat treatment at high temperature to improve the adhesion property of the coating. But this approach brings more problem than it solves. Due to mismatch of the thermal expansion coefficients between HAp coatings and substrates, large thermal stresses are generated during these processes. If and when such stresses are frozen in as residual stresses, it may lead to generation of cracks at the interface. The presence of such interfacial cracks not only decreases the coating adhesion, but also leads to coating delamination [39–42].

3.3.3 Issues Related to MIPS Processes

In very recent approaches the MIPS process [8–18] has also gained popularity. There are reasons for the same. Some of its major advantages are phase purity, high crystallinity, and adequate porosity for better osseointegration, along with a reasonably high and acceptable magnitude of mechanical properties. The other unique feature of the MIPS process is that it can deposit HAp on a variety of complex shapes. Moreover, it requires much less energy. This process is therefore much greener than the MAPS process, which is much more energy-intensive, and hence more cost-intensive.

3.4 Microplasma and Macroplasma Sprayed HAp Coatings: Pros and Cons

The practical challenges of the biomedical application of a conventional dense high-power macroplasma sprayed or MAPS-HAp coating include:

1. Occlusion of the porous surface
2. Too fast bioresorption
3. Late delamination with formation of particulate debris [43, 44]

The last factor is an issue of significant importance in the case of, e.g., hip implants because particulate debris of HAp might well accelerate polyethylene wear-induced granulomatous tissue response with an associated bone lysis [45]. On the other hand, the microplasma sprayed (MIPS) HAp coatings [8–14] possess a high degree (>90%) of crystallinity and a high porosity level (10–20%), which takes care of the controlled bioresorption and osseoconductivity [2, 40–42], i.e., the capability of supporting the ingrowth of sprouting capillaries, perivascular tissues, and osteoprogenitor cells from the recipient host bed into the three-dimensional structure of an implant or graft.

Very recently, the present author and coworkers have illustrated that MIPS-HAp coatings can also act as a promising coating method for implants in orthopaedic applications [8–14]. For instance, a MIPS-HAp-coated intramedullary pinning utilized for bone fracture fixation showed promising tissue response [9]. This could be achieved because compared to macroplasma (MAPS) sprayed HAp coatings, the MIPS process:

1. Requires much less plasmatron power (e.g., 1–4 kW cf. 10–40 kW or higher)
2. Generally avoids formation of impure and amorphous phases

3. Provides a much higher degree of phase purity and crystallinity (e.g., >80% cf. ≤70%)

4. Induces a relatively higher degree of porosity (e.g., ~20% cf. ≤2–10%) that facilitates bony tissue ingrowth

The term *micro* generally refers to the comparatively lower power level required for the conventional MAPS. However, too high levels of porosity are not really desirable in a thermally sprayed coating on implant material; although a high porosity might seem favorable to osteointegration, a very porous coating is likely to possess poor mechanical strength and poor adhesion to the substrate [2, 39, 40], so that the coating might delaminate or release highly undesirable debris, which obviously impairs osteointegration by eliciting an unfavorable tissue responses. Moreover, an excessive penetration of body fluids in the coating might cause dissolution of the coating material in the interface region, further weakening the adhesion strength. Therefore, it is generally recommended that a proper compromise between porosity and mechanical properties be devised [2, 12].

3.5 Influence of Plasma Spraying Parameters on HAp Coating

3.5.1 Role of Plasma Spraying Atmosphere, Spraying Current, and Stand-Off Distance

Yang and coworkers [46] showed that in the conventional atmospheric plasma sprayed (APS) HAp on the Ti-6Al-4V substrate, the coating characteristics, such as phase purity, crystallinity, morphology, and roughness, were affected by the plasma spraying parameters, especially the spraying atmosphere. For instance, they reported that Group I HAp coatings having H_2 as the secondary plasma gas contained higher impurity phases, less crystallinity, and better melting of the coating. In contrast, Group II HAp coatings having He as the secondary plasma gas revealed lower impurity levels, higher crystallinity, and worse melting of the coating.

Further, the amount of impurity phases in Group I HAp coatings [46] increased with H_2 content and spraying current, but was not related to the stand-off distance. The index of crystallinity for Group I HAp coatings decreased with increasing H_2 content, spraying current, and stand-off distance. However, for the Group II HAp coatings, the index of crystallinity was more than 50%, and almost 95% apatite with barely detectable (<5%) extra phases was obtained. They [46] opined that the simultaneous achievement of both better melting of the coating, e.g., as in the Group I HAp coatings, and a higher phase purity, e.g., as in the Group II HAp coatings, was difficult.

Because the heat content of diatomic H_2 gas was higher than that of the monoatomic He gas [47], the higher heat content in the plasma of Group I provided a greater ability to melt the starting HAp, leading to more opportunities for phase decomposition than possibly those in the other plasma of Group II. As such, the plasma spraying process subjects the HAp material to high temperatures with abundant heat content. Therefore, it is expected that the phase stability of HAp might no longer hold out perfectly. As a result, the HAp phase tends to decompose.

As per phase diagram and related literature data, the phase decomposition probably proceeds in several steps: At high temperature, possibly above 1450°C, there is decomposition of HAp to form α-TCP (α-$Ca_4(PO_4)_2$) and TP ($Ca_4P_2O_9$) [48]. Being an unstable phase at room temperature, α-TCP naturally transforms to β-TCP (β-$Ca_4(PO_4)_2$) (stable phase at room temperature) at about 1100°C. If the environment involves a high heat content, the TP phase would further decompose to form HAp and CaO [48]. Thus, the distinctive difference in phase composition between Group I and Group II HAp coatings is expected to happen [46].

As a result, higher amounts of impurity phases were obtained for HAp coatings sprayed using Group I plasma spraying parameters. In the same way, the impurity phase amounts increased with H_2 content and spraying current, owing to increase in the heat content in the plasma spraying parameters. In contrast, the high phase purity of the Group II HAp coatings, revealing lower impurity phase values, was due to the lower heat content of the plasma spraying parameters involved. It was further suggested [46] that since the heat content of the plasma was not changed with variation of the stand-off distance, the impurity phase contents of HAp coatings were not influenced by the stand-off distance variable [46].

3.5.2 Role of Gun Traverse Speed

For the HAp/Ti-6Al-4V composite coating, Quek et al. [49] indicated that the faster gun transverse speed results in a highly porous coating with low crystallinity and a mechanically weak bonding. In the other case, Yang and Chang [50] showed that for each kind of HAp coating, the substrate temperature of the specimens decreased with increasing surface speed of the gun transverse.

3.5.3 Role of Specimen Holder Arrangements

Owing to the higher surface speed and the corresponding better surface cooling, the specimens manufactured using the rotational holder (e.g., B-HAp coatings) showed [50] in general a lower substrate temperature than that obtained in a fixed holder (A-HAp coatings). The bonding strengths of the A-HAp coatings (9–16 MPa) were reported to be substantially lower than those of the B-HAp coatings (21–26 MPa). The data have been explained in terms of the role of porosity in the microstructure. The denser microstructure in B-HAp coatings gave better bonding strength, while the less dense microstructure in A-HAp coatings gave a comparatively poorer bonding strength [50].

3.5.4 Macroplasma Spraying vs. Microplasma Spraying

Recently Zhao et al. [51] showed that the microplasma sprayed coatings could be deposited with HAp content exceeding 85%. Just as during MAPS, during MIPS process also the electric current and spray distance both significantly influenced the coating structure and crystallinity. In addition, they found that the argon flow rate was also a very important parameter for the MIPS process. Recently, the present authors and coworkers reported the crystallinity of MIPS-coated HAp as ~80%, and further improvement (e.g., ~90%) was possible after post-annealing [8–14]. The porosity was also reported to be very high, e.g., about 11–20%, with adequate bonding strength [8–14].

3.6 Nanostructured HAp Coating

If we consider high-temperature thermally sprayed coatings, strictly speaking, they are not nanostructured, but rather micron scale. To obtain the nano-HAp coatings, chemical routes like the sol-gel dip method, biomimetics, and EPD are suitable methods. Further, physical vapor deposition, in particular the sputtering technique, and chemical vapor deposition are applied to develop nanostructured HAp. However, it is very difficult to develop the desired thickness, in the range of about 100 to 250 μm, especially for biomedical and orthopedic applications, by the aforesaid deposition techniques. An adequate porosity level is also difficult to achieve with these coating techniques. The plasma spray process with a starting HAp powder nanostructured particle made by the spray drying technique can be an alternative option to get nanostructures with adequate porosity level and thickness. A detailed discussion of this process is provided in Chapter 12.

3.7 HAp Composite Coating

It is a well-known fact that HAp possesses poor mechanical properties. During in vivo implantation, the structural integrity of the coating is one of the key factors, besides its favorable biological responses. To increase the mechanical properties, particularly the fracture toughness of the HAp coating, tough YSZ, TiO_2, and carbon nanotube (CNT) have been introduced as a second phase with the HAp [52–54]. The HAp-YSZ [52] and HAp-CNT composites [54] were developed by the MAPS method, whereas HAp-TiO_2 composites [53] were developed by the high-velocity oxy fuel (HVOF) technique. Apart from the HAp-based composites, the doping of fluorine by the dip coating technique can also improve the bonding strength of the HAp coating.

3.8 Plasma-Sprayed HAp Coating: Current Research Scenario

3.8.1 Robot-Assisted Plasma Spraying

Generally, for commercial development of HAp coating, the atmospheric or air plasma spraying method with a power level of 10–40 kW or more has been utilized. Nowadays, the uses of a manipulator and robotic arm have been introduced to improve the uniformity of the coatings. With the state-of-the-art robotic arm-controlled plasma gun, it is possible today to coat any contour- or complex-shaped component/implants with HAp.

3.8.2 Vacuum Plasma Spraying

Instead of an air environment, a vacuum environment is also popular for the plasma sprayed HAp coating's development. The advantage of the vacuum atmosphere is that it reduces the chance of oxidation and preserves a more hygienic way of coating preparation. Actually, first the vacuum plasma spray (VPS) chamber is evacuated in the range of about 0.1–0.08 mbar. Then the chamber is filled with inert gas at a low pressure of about 100 mbar. The plasma spraying follows as the next step and usually leads to a coating density better than that which could be attained with a conventional APS or MAPS process.

The VPS-HAp coatings were found to possess a lower residual stress, a higher crystallinity, and a higher level of porosity [55]. The effect of power levels on characteristics of VPS-HAp coatings had been studied by many researchers [56, 57]. Further, the efficacy of the VPS-HAp coatings showed the proof of stability in both SBF and fetal calf serum (FCS) mediums [58]. The osseointegration property of the VPS-HAp-coated implants were also successfully studied by Aebli et al. in sheep models [59]. They found a significant improvement of interfacial shear strength measured by the pull-out tests over all the time intervals, e.g., 1, 2, and 4 weeks after implantation.

3.8.3 Liquid Precursor Plasma Spraying

Development of both dense and porous nanostructured HAp coatings can also be produced utilizing liquid precursors, as mentioned in Chapter 2, by the liquid precursor plasma spraying (LPPS) technique [60, 61]. A much higher porosity (48%) and the pore size of 10–200 µm can be achieved in coatings deposited by the LPPS process. When evaluated with human osteo-blastic cell MG-63, the osteoblastic cell responses of such LPPS HAp coatings had also shown promising results.

3.8.4 Cold Spraying

The cold spraying technique falls in a wide category of thermal spraying. For deposition of metallic or ceramic coatings at lower temperatures, this method has become very popular nowadays. As mentioned in Chapter 2, in this process, a hot gas, e.g., He, Ni, or dry air at very high velocity (e.g., around 300–1200 m.s^{-1}), is used to propel either metallic or ceramic powder particles onto the substrates to affect deposition of a uniform coating. The coating basically forms due to the impact of the powder particles at super-sonic velocity rather than due to high-temperature melting. Thus, in this process, oxidation of substrates, poor crystallinity, introduction of residual stresses, etc., can be avoided [62]. Recently, Lu et al. [63] demonstrated the deposition of HAp-Ti composite coatings by the cold spraying method. The coatings also exhibited good bioactivity in SBF medium. In another recent report, Noorakma et al. [64] developed 20–30 µm thick HAp coatings on bio-degradable AZ51 magnesium alloys. They utilized a modified cold spraying technique. Here, instead of a higher temperature, i.e., 700–800°C, a hot gas at 400°C was utilized. They also studied the efficacy of the deposited HAp coatings in SBF medium. However, due to the low deforming ability or less ductility of HAp, as expected for a brittle material, the amount or quantum of works reported for HAp coatings deposited by the cold spraying tech-nique is rather limited.

3.8.5 Microplasma Spraying

To develop a phase-pure and porous, yet highly crystalline HAp coating, as mentioned in Chapters 1 and 2, plasma spraying with a very low plas-matron power (e.g., 1–2 kW, i.e., microplasma spraying) has recently been introduced by the present authors and coworkers [8–14], as well as other researchers [15–18]. In fact, from the reports of Dey and his coworkers [8–14], a few things have been proved. These are as follows [8–14]. The MIPS-HAp coating on SS316L metallic implants can be successfully and reproducibly deposited. These MIPS-HAp coatings possess a high degree of crystallinity (80–92%) and porosity (11–20%). The presence of porosity can be optimized in such a way as not to compromise the mechanical properties of the MIPS-HAp coatings. These coatings had high bonding strength (~20–13 MPa), high hardness (~4.5 GPa), large Young's modulus (~80 GPa), and high toughness (~0.6 MPa.m$^{0.5}$) with a unique damage-tolerant microstructure. In fact, these MIPS-HAp coatings showed a satisfactory result in in vitro SBF solution as well as in in vivo animal trial. On the other hand, Junker et al. [15, 16] have also established the efficacy of MIPS-HAp coatings in in vivo animal trials through goat and dog models.

3.9 Summary

In the present chapter, the HAp coating and its applications have been discussed in detail. The influence of plasma spraying parameters on HAp coating's characteristics has been elaborated. Further, the relative merits and demerits of high-temperature- and low-temperature-assisted HAp coating deposition processes have been critically analyzed. The state-of-the-art scenarios of the nanostructured HAp coatings, HAp composite coatings, and doped HAp-coatings have been described, with an analysis of their relative limitations and advantages. Of course, a major emphasis has been given on the relative efficacies of the HAp coatings deposited by the MAPS and MIPS processes. This discussion highlighted the major success of MIPS-HAp coatings developed in recent times.

Structural and chemical properties of HAp coatings will be discussed in Chapter 4. The emphasis will be on evaluation of different microstructural parameters, such as the phase purity, presence of multiple phases and crystallinity, porosity, stoichiometry, variation of thickness, spectroscopic investigation of surface groups, etc. Such discussions that include HAp coatings developed by different processes will be elaborated. Further, detailed microstructural features of MIPS-HAp coatings, such as the splat geometry and dimension, macro- and micropore sizes and their distributions, macro- and microcrack sizes and their distributions, etc., will be presented. The formation mechanisms of the highly porous, heterogeneous microstructure of the MIPS-HAp coatings shall be explained. In addition, the porosity dependencies of mechanical properties of MIPS-HAp will be discussed in Chapter 4, and the importance of pore shape in affecting the magnitudes of elastic constants in MIPS-HAp coatings will be critically evaluated.

References

1. Kweh S. W. K., Khor K. A., and Cheang P. 2000. Plasma-sprayed hydroxyapatite (HA) coatings with flame-spheroidized feedstock: microstructure and mechanical properties. *Biomaterials*, 21: 1223–1234.
2. Mancini C. E., Berndt C. C, Sun L., and Kucuk A. 2001. Porosity determinations in thermally sprayed hydroxyapatite coatings. *Journal of Materials Science*, 36: 3891–3896.
3. Chen Y., Zhang Q., Zhang T. H., Gan C. H., Zheng C. Y., and Yu G. 2006. Carbon nanotube reinforced hydroxyapatite composite coatings produced through laser surface alloying. *Carbon*, 44: 37–45.
4. Aria S. L., Mayor M. B., Pou J., Leng Y., Leon B., and Perez-Amora M. 2003. Micro- and nano-testing of calcium phosphate coatings produced by pulsed laser deposition. *Biomaterials*, 24: 3403–3408.
5. Gross K. A. and Samandari S. S. 2007. Nano-mechanical properties of hydroxyapatite coatings with a focus on the single solidified droplet. *Journal of Australian Ceramic Society*, 43: 98–101.
6. Khor K. A., Gu Y. W., Pan D., and Cheang P. 2004. Microstructure and mechanical properties of plasma sprayed HA/YSZ/Ti–6Al–4V composite coatings. *Biomaterials*, 25: 4009–4017.
7. Cheang P. and Khor K. A. 1996. Addressing processing problems associated with plasma spraying of hydroxyapatite coatings. *Biomaterials*, 17: 537–544.
8. Dey A. and Mukhopadhyay A. K. 2013. In-vitro dissolution, microstructural and mechanical characterizations of microplasma sprayed hydroxyapatite coating. *International Journal of Applied Ceramic Technology*, 11: 65–82.
9. Dey A., Nandi S. K., Kundu B., Kumar C., Mukherjee P., Roy S., Mukhopadhyay A. K., Sinha M. K., and Basu D. 2011. Evaluation of hydroxyapatite and β-tri calcium phosphate microplasma spray coated pin intra-medullarly for bone repair in a rabbit model. *Ceramics International*, 8: 1377–1391.
10. Dey A., Mukhopadhyay A. K., Gangadharan S., Sinha M. K., Basu D., and Bandyopadhyay N. R. 2009. Nanoindentation study of microplasma sprayed hydroxyapatite coating. *Ceramics International*, 35: 2295–2304.
11. Dey A., Mukhopadhyay A. K., Gangadharan S., Sinha M. K., and Basu D. 2009. Characterization of microplasma sprayed hydroxyapatite coating. *Journal of Thermal Spray Technology*, 18: 578–592.
12. Dey A., Mukhopadhyay A. K., Gangadharan S., Sinha M. K., and Basu D. 2009. Development of hydroxyapatite coating by microplasma spraying. *Materials and Manufacturing Processes*, 24: 1321–1330.
13. Dey A., Mukhopadhyay A. K., Gangadharan S., Sinha M. K., and Basu D. 2009. Weibull modulus of nano-hardness and elastic modulus of hydroxyapatite coating. *Journal of Materials Science*, 44: 4911–4918.
14. Dey A. and Mukhopadhyay A. K. 2010. Anisotropy in nano-hardness of microplasma sprayed hydroxyapatite coating. *Advances in Applied Ceramics*, 109: 346–354.
15. Junker R., Manders P. J. D., Wolke J., Borisov Y., and Jansen J. A. 2010. Bone-supportive behavior of microplasma-sprayed CaP-coated implants: mechanical and histological outcome in the goat. *Clinical Oral Implants Research*, 21: 189–200.

16. Junker R., Manders P. J. D., Wolke J., Borisov Y., and Jansen J. A. 2010. Bone reaction adjacent to microplasma sprayed CaP-coated oral implants subjected to occlusal load, an experimental study in the dog. Part I. Short-term results. *Clinical Oral Implants Research*, 21: 1251–1263.
17. He D., Zhao Q., Zhao L., and Sun X. 2007. Influence of microplasma spray parameters on the microstructure and crystallinity of hydroxyapatite coatings. *Chinese Journal of Materials Research*, 2007, 6, 659–663.
18. Zhao Q., He D., Zhao L., and Li X. 2011. In-vitro study of microplasma sprayed hydroxyapatite coatings in Hanks balanced salt solution. *Materials and Manufacturing Processes*, 26: 175–180.
19. Mohammadi Z., Moayyed A. A. Z., and Mesgar A. S. M. 2007. Adhesive and cohesive properties by indentation method of plasma-sprayed hydroxyapatite coatings. *Applied Surface Science*, 253: 4960–4965.
20. Gu Y. W., Khor K. A., and Cheang P. 2003. In vitro studies of plasma-sprayed hydroxyapatite/Ti-6Al-4V composite coatings in simulated body fluid (SBF). *Biomaterials*, 24: 1603–1611.
21. Garcia-Sanz F. J., Mayor M. B., Arias J. L., Pou J., Leon B., and Perez-Amor M. 1997. Hydroxyapatite coatings: a comparative study between plasma-spray and pulsed laser deposition techniques. *Journal of Materials Science: Materials in Medicine*, 8: 861–865.
22. Cheng G. J., Pirzada D., Cai M., Mohanty P., and Bandyopadhyay A. 2005. Bioceramic coating of hydroxyapatite on titanium substrate with Nd-YAG laser. *Materials Science and Engineering C*, 25: 541–547.
23. Chen Y., Zhang Y. Q., Zhang T. H., Gan C. H., Zheng C. Y., and Yu G. 2006. Carbon nanotube reinforced hydroxyapatite composite coatings produced through laser surface alloying. *Carbon*, 44: 37–45.
24. Arias J. L., Mayor M. B., Pou J., Leng Y., Leon B., and Perez-Amora M. 2003. Micro- and nano-testing of calcium phosphate coatings produced by pulsed laser deposition. *Biomaterials*, 24: 3403–3408.
25. Khor K. A., Li H., and Cheang P. 2003. Characterization of the bone-like apatite precipitated on high velocity oxy-fuel (HVOF) sprayed calcium phosphate deposits. *Biomaterials*, 24: 769–775.
26. Khor K. A., Gu Y. W., Quek C. H., and Cheang P. 2003. Plasma spraying of functionally graded hydroxyapatite/Ti-6Al-4V coatings. *Surface and Coatings Technology*, 168: 195–201.
27. Gross K. A. and Samandari S. S. 2009. Nanoindentation on the surface of thermally sprayed coatings. *Surface and Coatings Technology*, 203: 3516–3520.
28. Nieh T. G., Jankowsk A. F., and Koike J. 2001. Processing and characterization of hydroxyapatite coatings on titanium produced by magnetron sputtering. *Journal of Materials Research*, 16: 3238–3245.
29. Nieh T. G., Choi B. W., and Jankowski A. F. 2001. Synthesis and characterization of porous hydroxyapatite and hydroxyapatite coatings. Report submitted to Minerals, Metals and Materials Society Annual Meeting and Exhibition, Los Angeles.
30. Guo X., Gough J., and Xiao P. 2007. Electrophoretic deposition of hydroxyapatite coating on Fecralloy and analysis of human osteoblastic cellular response. *Journal of Biomedical Materials Research A*, 80: 24–33.

31. Ma J., Wang C., and Peng K. W. 2003. Electrophoretic deposition of porous hydroxyapatite scaffold. *Biomaterials*, 24: 3505–3510.
32. Liu D. M., Troczynski T., and Tseng W. J. 2001. Water-based sol-gel synthesis of hydroxyapatite: process development. *Biomaterials*, 22: 1721–1730.
33. Cheng K., Zhang S., Weng W., Khor K. A., Miao S., and Wang Y. 2008. The adhesion strength and residual stress of colloidal-sol gel derived [beta]-tricalcium-phosphate/fluoridated-hydroxyapatite biphasic coatings. *Thin Solid Films*, 516: 3251–3255.
34. Habibovic P., Barrere F., vanBlitterswijk C. A., de Groot K., and Layrolle P. 2002. Biomimetic hydroxyapatite coating on metal implants. *Journal of the American Ceramic Society*, 85: 517–522.
35. Chakraborty J., Sinha M. K., and Basu D. 2007. Biomolecular template induced biomimetic coating of hydroxyapatite on an SS316L substrate. *Journal of the American Ceramic Society*, 90: 1258–1261.
36. Li X., Huang J., and Edirisinghe M. 2008. Development of nano-hydroxyapatite coating by electrohydrodynamic atomization spraying. *Journal of Materials Science: Materials in Medicine*, 19: 1545–1551.
37. Huang J. S., Jayasinghe N. S., Best M., Edirisinghe M. J., Brooks R. A., and Bonfield W. 2004. Electrospraying of a nano-hydroxyapatite suspension. *Journal of Materials Science*, 39: 1029–1032.
38. Kim H. W., Yoon B. H., Koh Y. H., and Kim H. E. 2006. Processing and performance of hydroxyapatite/fluorapatite double layer coating on zirconia by the powder slurry method. *Journal of the American Ceramic Society*, 89: 2466–2472.
39. Pawlowski L. 1995. *The science and engineering of thermal spray coatings.* Chichester: John Wiley & Sons.
40. Gross K. A. and Berndt C. C. 2002. Biomedical application of apatites. *Reviews in Mineralogy and Geochemistry*, 48: 631–672.
41. Heimann R. B. 2006. Thermal spraying of biomaterials. *Surface and Coatings Technology*, 201: 2012–2019.
42. Hench L. L. and Ethridge E. C. 1982. *Biomaterials: an interfacial approach.* New York: Academic Press.
43. Bauer T. W., Geesink R. C., Zimmerman R., and McMahon J. T. 1991. Hydroxyapatite-coated femoral stems. Histological analysis of components retrieved at autopsy. *Journal of Bone and Joint Surgery*, 73: 1439–1452.
44. Collier J. P., Surprenant V. A., Mayor M. B., Wrona M., Jensen R. E., and Surprenant H. P. 1993. Loss of hydroxyapatite coating on retrieved total hip components. *Journal of Arthroplasty*, 8: 389–393.
45. Rothman R. H., Hozack W. J., Ranawat A., and Moriarty L. 1996. Hydroxyapatite-coated femoral stems. A matched-pair analysis of coated and uncoated implants. *Journal of Bone and Joint Surgery*, 78: 319–324.
46. Yang C. Y., Wang C., Chang E., and Wu J. D. 1995. The influences of plasma spraying parameters on the characteristics of hydroxyapatite coatings: a quantitative study. *Journal of Materials Science: Materials in Medicine*, 6: 249–257.
47. Ingham H. S. and Shephard A. P. 1965. *Flame spray handbook.* New York: Metco, 11.
48. de Groot K., Klein C. P. A. T., Wolke J. G. C., and de Blieck-Hogervorst J. M. A. 1990. Plasma-sprayed coatings of calcium phosphate. In *CRC handbook of bioactive ceramics*, Yamamuro T., Hench L. L., and Wilson J. (Eds.). Boca Raton, FL: CRC Press, 3–16.

49. Quek C. H., Khor K. A., and Cheang P. 1999. Influence of processing parameters in the plasma spraying of hydroxyapatite/Ti-6Al-4V composite coatings. *Journal of Materials Processing Technology*, 89–90: 550–555.
50. Yang Y. C. and Chang E. 2003. The bonding of plasma-sprayed hydroxyapatite coatings to titanium: effect of processing, porosity and residual stress. *Thin Solid Films*, 444: 260–275.
51. Zhao L., Bobzin K., Ernst F., Zwick J., and Lugscheider E. 2006. Study on the influence of plasma spray processes and spray parameters on the structure and crystallinity of hydroxylapatite coatings. *Materialwissenschaft und Werkstofftechnik*, 37: 516–520.
52. Fu L., Khor K. A., and Lim J. P. 2002. Effects of yttria-stabilized zirconia on plasma-sprayed hydroxyapatite/yttria-stabilized zirconia composite coatings. *Journal of the American Ceramic Society*, 85: 800–806.
53. Li H., Khor K. A., and Cheang P. 2002. Titanium dioxide reinforced hydroxyapatite coatings deposited by high velocity oxy-fuel (HVOF) spray. *Biomaterials*, 23: 85–91.
54. Balani K., Anderson R., Laha T., Andara M., Tercero J., Crumpler E., and Agarwal A. 2007. Plasma-sprayed carbon nanotube reinforced hydroxyapatite coatings and their interaction with human osteoblasts in vitro. *Biomaterials*, 28: 618–624.
55. Gledhill H. C., Turner I. G., and Doyle C. 1999. Direct morphological comparison of vacuum plasma sprayed and detonation gun sprayed hydroxyapatite coatings for orthopaedic applications. *Biomaterials*, 20: 315–322.
56. Chang C., Shi J., Huang J., Hu Z., and Ding C. 1998. Effects of power level on characteristics of vacuum plasma sprayed hydroxyapatite coating. *Journal of Thermal Spray Technology*, 1998, 7: 484–488.
57. Chang C.-K., Shi J.-M., Hu Z.-Y., Huang J.-Q., and Ding C.-X. 1998. Effect of spray power on characteristics of vacuum plasma sprayed hydroxyapatite coatings. *Journal of Inorganic Materials*, 13: 219–224.
58. Ha S.-W., Reber R., Eckert K.-L., Petitmermet M., Mayer J., Wintermantel E., Baerlocher C., and Gruner H. 1998. Chemical and morphological changes of vacuum-plasma-sprayed hydroxyapatite coatings during immersion in simulated physiological solutions. *Journal of the American Ceramic Society*, 81: 81–88.
59. Aebli N., Krebs J., Stich H., Schawalder P., Walton M., Schwenke D., Gruner H., Gasser B., and Theis J.-C. 2003. In vivo comparison of the osseointegration of vacuum plasma sprayed titanium- and hydroxyapatite-coated implants. *Journal of Biomedical Materials Research Part A*, 66A: 356–363.
60. Huang Y., Song L., Liu X., Xiao Y., Wu Y., Chen J., Wu F., and Gu Z. 2010. Hydroxyapatite coatings deposited by liquid precursor plasma spraying: controlled dense and porous microstructures and osteoblastic cell responses. *Biofabrication*, 2: 045003.
61. Huang Y., Song L., Liu X., Xiao Y., Wu Y., Chen J., Wu F., and Gu Z. 2010. Characterization and formation mechanism of nano-structured hydroxyapatite coatings deposited by the liquid precursor plasma spraying process. *Biomedical Materials*, 5: 054113.
62. Singh H., Sidhu T. S., and Kalsi S. B. S. 2012. Cold spray technology: future of coating deposition processes. *Frattura ed Integrità Strutturale*, 22: 69–84.

63. Lu L., Zhou X., and Mohanty P. 2014. In vitro immersion behavior of cold sprayed hydroxyapatite/titanium composite coatings. *Journal of Biosciences and Medicines*, 2: 10–16.
64. Noorakma A. C. W., Zuhailawati H., Aishvarya V., and Dhindaw B. K. 2013. Hydroxyapatite-coated magnesium-based biodegradable alloy: cold spray deposition and simulated body fluid studies. *Journal of Materials Engineering and Performance*, 22: 2997–3004.

4

Structural and Chemical Properties of Hydroxyapatite Coating

4.1 Introduction

In this chapter we discussed the structural and chemical properties of HAp coating. We provide information on the phase analysis, stoichiometry and Ca/P ratio, microstructure, surface roughness, etc. The data presented in Table 4.1 [1–20] summarize the data from the literature on degree of crystallinity, thickness, and porosity of HAp coatings deposited by macroplasma spraying (MAPS) and other methods.

4.1.1 Thickness of HAp Coatings

The data presented in Table 4.1 show that depositions of both thin (3–23 μm) and thick HAp-based coatings have been reported in the literature. The relatively thinner coatings were deposited by low-temperature processes like sol-gel and biomimetic routes [13, 15–17]. The limitations of these processes are that generally, thicker coatings cannot be deposited by these techniques. In fact, even if an attempt is made to deposit thicker coatings by such techniques, often the coatings delaminate due to poor adhesion. That is why coatings of higher thickness, e.g., 50–300 μm, are usually deposited by the high-temperature processes, like MAPS or microplasma spraying (MIPS) [1–12, 18–20]. In general, for the in vivo application, the most preferred coating thickness is at least more than 150 μm. The adhesion properties of both MAPS and MIPS coatings were comparatively much superior to those of the sol-gel and biomimetic HAp coatings. However, due to high amount of residual stress, often MAPS-HAp coatings showed long microcracks in directions both parallel to and perpendicular to the substrate, as shown in Figure 4.1a, b [2]. The major problem of concern is that such MAPS-HAp coatings may fail or delaminate at a later stage in in vivo implantation. In contrast, MIPS coatings didn't show any long cracks. It may be plausible that, as the input plasmatron power was less, so the amount of heat and temperature were

TABLE 4.1

Degree of Crystallinity (X_c), Thickness (t_c), and Porosity (p) of HAp Coatings

Coating Conditions					
P/S	**CP**	X_c (%)	t_c (μm)	p (%)	**Reference**
HAp/Ti-6Al-4V	MAPS		180 ± 20	3.5–4	[1]
HAp/Ti-6Al-4V	MAPS	27–47	180 ± 20	6–9	[2]
HAp/Ti-6Al-4V	MAPS	44–73	180 ± 20		[3]
HAp/Ti-6Al-4V	MAPS		200		[4]
HAp/Ti-6Al-4V	MAPS	50–80	200	3.5–9	[5]
HAp/Ti-6Al-4V	MAPS	22–64	200	4–12	[6]
HAp/Ti-6Al-4V	MAPS		50–200		[7]
HAp/Ti	MAPS		200		[8]
HAp composite/Ti-6Al-4V	MAPS				[9]
HAp-BG/Ti-6Al-4V	MAPS		100		[10]
HAp-Na- Ti/Ti-6Al-4V	MAPS		300		[11]
HAp, HAp-Ti-6Al-4V/Ti-6Al-4V	MAPS			16–18	[12]
YSZ-HAp/SS	SG		23		[13]
HAp/Ti-6Al-4V	SG				[14]
HAp/Ti-6Al-4V	Aero SG		5		[15]
β-TCP, FHAp/Ti-6Al-4V	SG		~3		[16]
HAp/BG, GC, H_d, Ti	B		10		[17]
HAp/SS316L	MIPS	80, 91	~200	20–11	[18–20]

Note: P/S = coating/substrate, CP = method of coating, X_c = degree of crystallinity of the coating, t_c = thickness of the coating, MAPS = macroplasma spraying, MIPS = microplasma spraying technique, SG = sol-gel technique, B =biomimetic process, p = percentage of porosity, FHAp = fluoridated hydroxyapatite, BG = bioglass, GC = glass-ceramic, and H_d = dense hydroxyaptite.

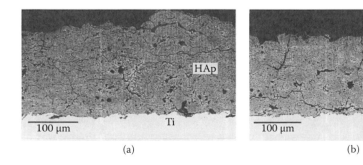

(a)   (b)

FIGURE 4.1
The MAPS-HAp coatings showed (a) parallel and (b) perpendicular long microcracks with respect to the substrate. (Reprinted from Yang and Chang, *Thin Solid Films*, 444: 260–275, 2003. With permission from Elsevier.)

lesser compared to those in the case of the MAPS-HAp coatings. Hence, the amount of residual stress was possibly smaller than that generated in the conventional MAPS-HAp coatings. A detailed discussion on the microstructure development of MIPS-HAp coatings will be taken up later in this chapter.

4.1.2 Porosity of HAp Coatings

The MAPS-HAp coatings are dense, and hence barring one or two exceptions where porosity greater than 15% has been reported [12], they are generally restricted to the small range of 3.5 to 12% [1, 2, 5, 6]. The porosity is usually measured by the image analysis technique from a large number of scanning electron microscopy (SEM) or field emission scanning electron microscopy (FESEM) photomicrographs, which adequately represent the microstructure of a given coating. The typical dense microstructure of a MAPS-HAp coating deposited in-house on SS316L is shown in Figure 4.2. Now, as we have already discussed in Chapters 2 and 3, the MIPS-HAp coatings always show porosity, e.g., ~20% [18–20] or more, that is obviously much higher than what could be achieved with the conventional MAPS-HAp coatings. Thus, it needs to be reemphasized once more here that from the in vivo application point of view, the efficacy of osseointegration for a porous HAp coating is much better than that of the denser one.

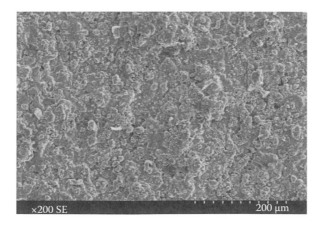

FIGURE 4.2
Dense microstructure of MAPS-HAp coatings.

4.1.3 Crystallinity of HAp Coatings

The degree of crystallinity of the MAPS coatings as reported in the literature covers a truly spectacular range, i.e., as low as 22% and as high as 80% [2, 3, 5, 6]. Post-deposition annealing hikes the degree of crystallinity further. The impurity phases like tricalcium phosphate (TCP), tetracalcium phosphate (TTCP), calcium oxide (CaO), etc., are generally always present in MAPS coatings, e.g., as shown in Figure 4.2. The degree of crystallinity of this particular MAPS coating developed in-house on SS316 was ~53%. The degree of crystallinity (X_c) corresponding to the fraction of crystalline phase present in the examined volume was evaluated from the x-ray diffraction (XRD) data using following relationship [21]:

$$X_c \approx 1 - (V_{112/300}/I_{300}) \tag{4.1}$$

where I_{300} is the intensity of (300) reflection and $V_{112/300}$ is the intensity of the hollow between (112) and (300) reflections, which completely disappears in noncrystalline samples. Just to give a typical illustration of how things are done, the XRD pattern of an in-house-developed MAPS-HAp coating is shown in Figure 4.3, and the enlarged view of the peaks corresponding to the (211), (112), and (300) planes referred to in Equation (4.1) is shown in Figure 4.4.

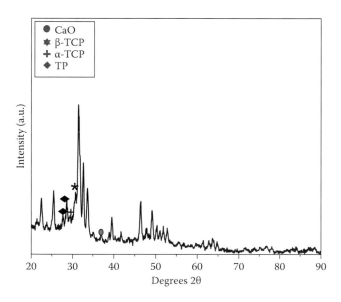

FIGURE 4.3
XRD pattern of MAPS-HAp coating show several impurity phases.

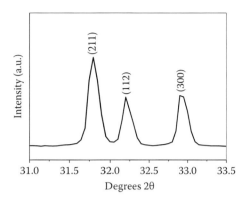

FIGURE 4.4
Enlarged view of a typical XRD pattern of MAPS-HAp to identify (211), (112), and (300) planes.

4.2 Stoichiometry of HAp

The data on chemical analysis as obtained from inductively coupled plasma atomic emission spectroscopy (ICP-AES) analysis of the HAp powder prepared by the wet chemical route [18–20] are given in Table 4.2. The Ca/P molar ratio was calculated to be ~1.67, indicating that the powder was stoichiometric in nature. Further, from transmission electron microscopy (TEM) based EDX data, the Ca/P ratio of the HAp powder was found as ~1.67. These data tallied also with the Ca/P ratio of the powder as measured by the ICP-AES technique, as mentioned earlier. Further, the experimentally measured Ca/P ratio of the MIPS-HAp coatings was exactly the same as that of the HAp powder.

TABLE 4.2

Chemical Analysis of HAp Powder

Constituents	Wt%
CaO	53.58
P_2O_5	40.89
Fe_2O_3	0.02
MgO	0.5
Pb	Trace
Cd	Trace

Source: Dey et al., *Materials and Manufacturing Processes*, 24: 1321–1330, 2009. Reprinted with permission from Taylor and Francis Group.

4.3 Phase Analysis of MIPS-HAp Coatings

XRD patterns of the HAp coating in the as-sprayed condition and after heat treatment at 600°C are shown in Figure 4.5a, b [18–20]. The XRD pattern of the as-sprayed coating (Figure 4.5a) showed the major presence of the peaks corresponding to the HAp phase (JCPDS file 09-0432). However, in addition to the peaks corresponding to those of the pure HAp phase, only two other minor peaks of crystalline α-TCP and TTCP were identified. Similar observations were also reported by other researchers [6, 22]. The degree of crystallinity of the as-sprayed coating was calculated as ~80% (Figure 4.5a), which was slightly less than that of the HAp powder, e.g., ~90%. This was possible due to the presence of (a) crystalline phases other than the HAp phase, as mentioned earlier, or (b) amorphous phase, which could be formed during the microplasma spraying process itself. On the other hand, XRD pattern of the heat-treated coating (Figure 4.5b) showed all the peaks characteristic of the crystalline HAp phase only. Realistically, the degree of crystallinity was enhanced by 12% after post-heat treatment at 600°C. It had happened most likely due to the recrystallization of the amorphous phase that was perhaps present in the as-sprayed MIPS-HAp coating. It needs to be emphasized further that an increase of post-deposition heat treatment temperature to a magnitude higher than 600°C had already been reported to induce an adverse effect on both the substrate [23] and the coating crystallinity [24]. Therefore, in the research conducted by the present authors and coworkers, no attempt was made in the present work [18–20] to enhance the post-deposition heat treatment temperature to a magnitude higher than 600°C.

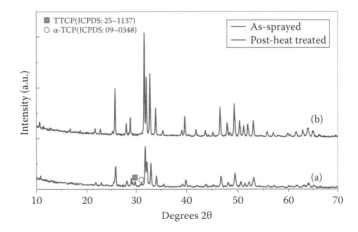

FIGURE 4.5
XRD pattern of HAp coating: (a) as sprayed and (b) post-heat treated at 600°C. (Reprinted from Dey et al., *Materials and Manufacturing Processes*, 24: 1321–1330, 2009. With permission from Taylor and Francis Group.)

4.4 Spectroscopic Investigation of MIPS-HAp Coatings

The data on Fourier transform infrared spectroscopy (FTIR) spectra of the HAp coating are shown in Figure 4.6a and b for both the as-sprayed coating and the coating post-heat treated at 600°C, respectively [18, 20]. FTIR spectra of both the as-sprayed coating and the coating post-heat treated at 600°C (Figure 4.2a, b) were comparable to those reported in the literature [25–30].

For instance, the band at around 961 cm^{-1} in the MIPS-HAp coating (962 cm^{-1}, Figure 4.6a) was the characteristic of nondegenerate symmetric stretching of the PO_4^{3-} group (v_1) in HAp [25]. The doubly degenerate O-P-O bending band (v_2) was present at 469 cm^{-1} in the as-sprayed coating (Figure 4.6a)

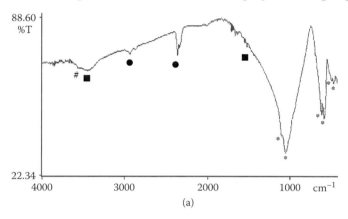

(a)

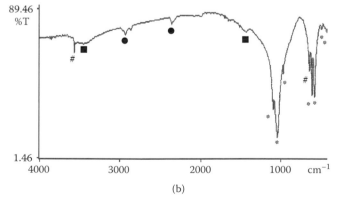

(b)

PO_4^{3-},# OH^-, ■ *Absorbed moisture and* ● *Absorbed CO$_2$*

FIGURE 4.6
FTIR spectra of the MIPS-HAp coating: (a) as sprayed and (b) post-heat treated at 600°C. (Reprinted from Dey et al., *Materials and Manufacturing Processes*, 24: 1321–1330, 2009. With permission from Taylor and Francis Group.)

and at 477 cm^{-1} (Figure 4.6b) in the post-heat-treated HAp coating. The bands at 1046 and 1092 cm^{-1} for both the as-sprayed coating (Figure 4.6a) and the post-heat-treated coating (Figure 4.6b) signify asymmetric O-P-O stretching band (v_3). The bands located at 569 and 600 cm^{-1} of the FTIR spectrum of the as-sprayed (Figure 4.6a) and post-heat-treated (Figure 4.6b) coatings stand for the triply degenerate asymmetric O-P-O bending band (v_4) [25].

It may be also noted from the data presented in Figure 4.6b that because there was high crystallinity in the post-heat-treated coating, the corresponding peaks at 569 and 600 cm^{-1} were very sharp and were split with a sharp bend. It was also interesting to note that the characteristic bending band for OH^{-1} occurred at 632.3 cm^{-1} in the HAp powder [18–20], but it disappeared in the as-sprayed coating (Figure 4.6a), thereby possibly suggesting that the amorphous phase in the as-sprayed coating would have very little, if any, hydroxyl group incorporated within its structure. However, post-heat treatment at 600°C led to the reappearance of the characteristic bending band for OH^{-1} at 631 cm^{-1} (Figure 4.6b), possibly implying the reestablishment of a hydroxylated HAp structure. Similar observations have also been reported by other researchers [26]. This reestablishment of a hydroxylated HAp structure may be linked to the increase in crystallinity (92%) of the HAp coating. The broader band in the as-sprayed coating (Figure 4.6a) at about 960 to 1100 cm^{-1} was split into two appreciable adsorption bands after post-heat treatment at 600°C (Figure 4.6b). Thus, the pronounced peaks at 1046 and 1092 cm^{-1} of the coating (Figure 4.6b) were linked to enhanced crystallinity [26].

However, some minor peaks, as reported by other workers [27–30], were also observed in the present work (Figure 4.6a, b). As mentioned earlier, their occurrence could possibly be attributed to the absorption of moisture and CO$_2$. Similar observations have been reported by other researchers also [27, 30]. There was a larger bulge on the right side of the small adsorption peak at 3569 cm^{-1} in the spectrum of as-sprayed coating (Figure 4.6a). After the post-heat treatment at 600°C, this bulge was substantially reduced and the adsorption peak became sharper (Figure 4.6b). The band at 3571 cm^{-1} has been referred to as a stretching band of OH^{-1} in the literature [26]. This information suggests that a rehydroxylation occurred in the MIPS-HAp coatings of the present work following the heat treatment. Thus, the overall observations of the FTIR spectra corroborated well with the XRD data.

4.5 Microstructure of MIPS-HAp Coating

4.5.1 As-Sprayed Condition

The surface morphology of the as-sprayed top surface, i.e., plan section, is shown through SEM (Figure 4.7a, b) [18, 20]. The microstructure of the

(a)   (b)

FIGURE 4.7
SEM images of the plan section of the MIPS-HAp coating: (a) as-sprayed, at low magnification, and (b) as-sprayed, at high magnification. Unmelted splats retaining a nonflattened core (white bold arrow), macro- and micropores (white bold arrowheads), intra- and intersplat cracks (black thin arrow), and deformed splats (black bold arrow). (Reprinted from Dey et al., *Materials and Manufacturing Processes*, 24: 1321–1330, 2009. With permission from Taylor and Francis Group.)

coating was heterogeneous and also highly porous. These open pores favor bony tissue ingrowth, as mentioned earlier. Others researchers [31–33] have also mentioned the heterogeneous structure of the HAp coatings.

The splat size of the coating was ~50–70 µm. The macro- and micropore sizes of the coating were ~10–50 µm and ~1 µm, respectively. The complex microstructure of the present coating consisted of completely or partially molten, deformed, and unmelted splats retaining a nonflattened core, micropores, macropores, intersplat, and intra-splat cracks (Figure 4.7b), and those are schematically illustrated in Figure 4.8 [34, 20].

Further, high-magnification SEM photomicrographs of the as-deposited MIPS-HAp coatings are shown in Figures 4.9a–h [18, 19]. Layer by layer stacking of molten splats and the interface between splats are depicted in

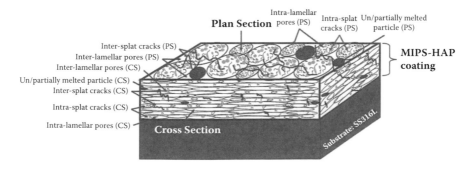

FIGURE 4.8 (See color insert.)
Schematic of the structure and nature of the MIPS-HAp coating. (Reprinted from Dey et al., *Journal of Materials Science*, 44: 4911–4918, 2009. With permission from Springer.)

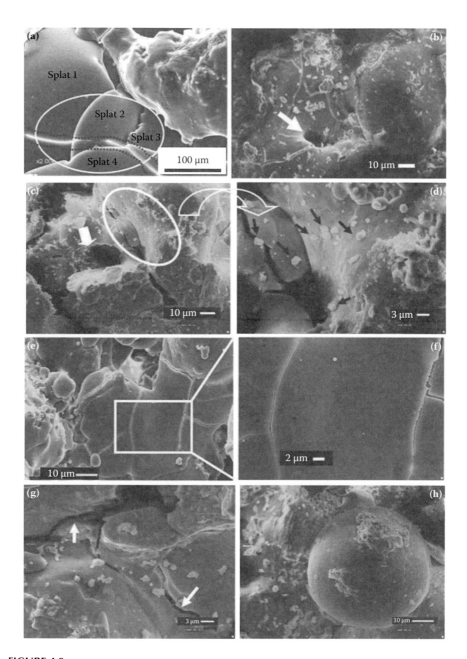

FIGURE 4.9
SEM images of as-sprayed microstructure of the MIPS-HAp coating: (a) stacking of splats, (b) macropore in a single splat, i.e., intrasplat pore, (c) macropore in between the splats, i.e., inter-splat pore, (d) tiny worn-out parts of splats, (e) cracks in a single splat, i.e., intrasplat cracks, (f) intrasplat cracks at higher magnification, (g) cracks in between splats, i.e., inter-splat cracks, and (h) an unmelted HAp granule. (Modified/reprinted from Dey et al., *Journal of Thermal Spray Technology*, 18: 578–592, 2009. With permission from Springer.)

Figure 4.9a. Further, a long intrasplat crack was found, indicated by the black dotted line in Figure 4.9a. In addition, typical intrasplat and intersplat macropores are shown in Figure 4.9b and c, respectively, while torn-out small parts of splats with different magnifications are also shown in Figure 4.9c, d. The presence of inclined cracks in between splats is shown at a lower magnification in Figure 4.9c and at a higher magnification in Figure 4.9d. These worn-out small parts were presumably produced due to impact between HAp granules and the substrate during the deposition process by the MIPS method, and consequently might have been surrounded at the coating surface.

Further, the cracks inside a given splat, i.e., typical intrasplat cracks, are shown in Figure 4.9e, f. The portion of intrasplat cracks marked in Figure 4.9e is shown at higher magnification in Figure 4.9f. It clearly shows also that the cracks grew both parallel and perpendicular to the microplasma spraying directions. Further, in the microstructure the cracks got deflected in an indiscriminate fashion. This was most likely governed by the local thermomechanical history during the microplasma spraying process. Similarly, there were cracks formed in between the splats, as shown in Figure 4.9g. The widths of intersplat cracks, as shown in Figure 4.9g, are larger than those of intrasplat cracks (Figure 4.9e, f). These cracks occurred probably due to the coefficient of thermal expansion (CTE) mismatch between the substrate SS316L metal and the deposited HAp ceramics. Further, unmelted spherical granule was also found in the MIPS-HAp coating microstructure (Figure 4.9h), which could also possibly be linked to hinder further improvement in the crystallinity of the HAp coating. All these factors, as discussed above, contribute thus to the formation of a very heterogeneous microstructure in the present MIPS-HAp coating, as also shown schematically in Figure 4.8.

4.5.2 Microstructure of MIPS-HAp Coatings in the Polished Condition

The polished top surface of the coating, i.e., plan section, is shown in Figure 4.10a–e at different regions and at various magnifications. The coating showed the characteristic presence of a large number of macrocracks, microcracks, cracks in between two splats, and cracks confined inside single splats, macro- and micropores, etc. The average volume percent open porosity as measured from image analysis of a multitude of the different SEM and FESEM micrographs of the coating was ~19.17 ± 1.98. A collection of the SEM images is shown in Figure 4.10e, f for the polished cross-sectional surfaces of MIPS-HAp coatings.

4.5.3 Splat Geometry and Dimension

The distribution of splat size and aspect ratio of the splats were as shown by the data presented in Figure 4.11a and b, respectively. The average splat size was ~64.28 ± 8.12 µm. The typical range of splat size was ~45–85 µm.

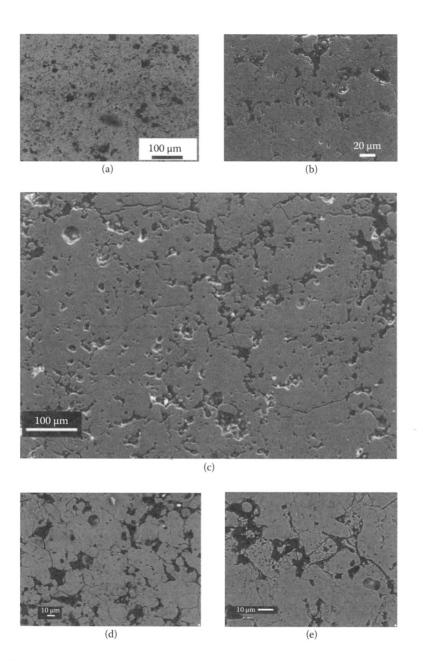

(a)   (b)

(c)

(d)   (e)

FIGURE 4.10
Polished plan section of the MIPS-HAp coating: (a) optical microscopy image, (b) SEM image, and FESEM images (c) at low magnification, showing the clear evidence of pores, microcracks, and other characteristic defects; (d) at higher magnification, showing pores and individual details of splats; and (e) at still higher magnification, showing details of microcracks and other defects. (Modified/reprinted from Dey et al., *Journal of Thermal Spray Technology*, 18: 578–592, 2009. With permission from Springer.)

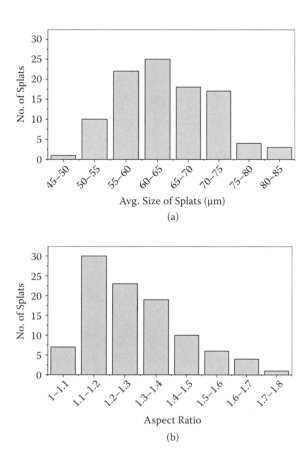

FIGURE 4.11

Number of splats as a function of (a) average size of splats and (b) aspect ratio. (Reprinted from Dey et al., *Journal of Thermal Spray Technology*, 18: 578–592, 2009. With permission from Springer.)

Similarly, the average aspect ratio of the splats was ~1.27 ± 0.15, and the typical range of aspect ratio was ~1–1.8. Thus, the shapes of the splats were mostly close to the shape of a small ellipsoid, as expected. The slight elongation along the major axis occurred because most of the splats finally assume a pancake shape [5].

4.5.4 Analysis of Splat Formation

It is indeed quite clear that the utilization of a low plasma power, together with the use of a quite coarse powder (d_{50} ~ 67 µm), resulted in the formation of a very large amount of unmelted material (Figure 4.5h). As already opined by Heimann [35], such unmelted material could well have had contributed to the high crystallinity of the coating because most of the molten material forms amorphous phases or TCP and TTCP when impact quenched. The same factor had possibly also contributed to the high porosity of the coating because as the partially unmelted particles do not flatten completely, they leave many voids in between them. Moreover, they cause a geometrical hindrance to the spreading of the other fully molten droplets, thereby leading to genesis of more open internal void spaces, and hence contributing further to the porosity aspect of the coating.

These facts suggested that the MIPS-HAp coatings were built up through freezing and flattening of granule. The freezing time of the HAp granules would be of the order of 0.2 to 1 μs. During this time about 50 to 100 granules may impinge upon a unit area, e.g., per m² of the substrate. As a result, the freezing of each individual HAp granule was most likely to be a completely isolated phenomenon. This implied that such freezing of a given molten granule would happen independent of the presence of other HAp granules. Thus, each granule would be frozen before the next HAp granule arrived. This is how possibly the lamellar structure of the MIPS-HAp coating had formed. The formation process of the coating may involve the following three steps:

1. The molten HAp granules impinge onto the SS316L substrate to form the splats.
2. These splats would spread parallel to the plane of the substrate as they form the individual lamellae.
3. These lamellae, stacked alongside each other and one on top of the other, form the microstructure of the MIPS-HAp coating of the present work.

The structures of individual lamellae would vary from each other because the wetting and flow properties would vary from one splat to another. This has to be a stochastically happening statistical event. Thus, the size of individual lamellae would also vary from each other because in reality, the HAp granule's temperature distribution would not be uniform throughout the volume, and a differential cooling rate was expected to prevail. Even the cooling rate would be affected by the extent of contact of the flattening HAp granule with the SS316L substrate, because the surface on which the HAp granules had flattened was very rough, with localized sharp asperities obtained by grit blasting. Further, once the first layer of HAp granules cover up the substrate, the second and subsequent layers of HAp granules would experience a completely different heat transfer scenario, because their access to the SS316L substrate would be rather limited. Therefore, the individual lamellae would have incongruent shape and structure. In other words, there would be gaps between the splats and the lamellae in the plane parallel to that of the substrate along the X-Y directions, and also along the thickness of the coating, as the MIPS-HAp coating would grow along the Z direction. This process would lead to the formation of a highly heterogeneous microstructure, as was indeed experimentally observed in the present MIPS-HAp coating.

The high porosity in the coating was formed because there could be less complete melting of the HAp granules due to lower inherent power input of the present MIPS process. Even if there were complete melting, due to lower power input, the temperature generated in the plasma would be much lower than that in a MAPS process, and as a result, the molten droplets could have had a much higher viscosity and a corresponding decrease in the flow properties. This would lead to a decrease in the flattening ratio of the spreading splats. The splat size was ~50–70 μm, i.e., very similar to the average size of the original powder, although the flattening of a molten droplet should result in a splat diameter that is (at least) three to four times larger than the original droplet diameter. Two explanations are possible: on the one hand, a very low splat flattening degree is caused by the retention of a large portion of unmelted material within most of the sprayed particles; on the other hand, it is likely that most of the largest particles are almost completely unmelted and rebound without deposition, because of the very low plasma power, so that most of the coating is originated by the finer fraction of the powder. However, this would also imply very low deposition efficiency.

Further, since the plasma spot size was small, e.g., about 3–5 mm, and the sinter granule size was much larger at about 67 μm, the number of such splats being formed per unit time per unit area would not be as high as in the case of a MAPS process. Thus, the area coverage of the coating would leave behind a lot of isolated islands, as it would grow across the thickness. Given the fact that there would be more incongruence than congruence in the shape and size of the individual lamellae, which are the basic building blocks of the coating, it was expected that there would be a high volume percent of open porosity available in such MIPS-HAp coatings. Such a picture rationalized the experimental observation of about 20 vol% open porosity in the coating.

4.5.5 Why Are Micropores Formed?

The data indicate that the macropore size and micropore size of the coating were ~10–50 μm and ~1 μm, respectively. It is possible that the micropores in the coating were formed during the MIPS process due, to some extent, to a solidification contraction process and a splat filling mechanism. It should also be borne in mind that the starting HAp granules had interparticle pores of the order of a micron, and as a result of this, it may be alternatively argued that the micropores of the size of a micron or so were already existing as isolated pores within the starting HAp granules [31].

4.5.6 How Are Macropores Formed?

There are two conceivable ways in which the macropores could have been formed. One such means is the simple geometric gap that remains unfilled due to a local, yet large amount of incongruence in the shape and size of the individual lamellae that, during the MIPS process, were forced to sit side by side and also on top of one another. The other possibility could be that there were unmelted granule cores that were pulled out during the grinding and polishing processes. According to the theory of the dynamics of gas bubbles within a given liquid, small gas bubbles are expected to coalesce to form a large bubble within a droplet. Accordingly, researchers have opined that in-flight clustering of the micropores might also provide an additional mechanism for the formation of macropores [31]. But even given the possibility of this scenario being true, it still remains very interesting indeed to note that even the largest macropore size (~50 μm) was less than that of the HAp granule size (~67 μm).

4.5.7 Coating Cross Section

How good or how bad is the cross section of the MIPS-HAp coating? To answer this question, at least 10 SEM or FESEM images (Figure 4.12a–f) were taken at each of the five randomly picked up locations of the coating. Then the characteristic features, e.g., micropore size, macrpore size, and microcrack length, were analyzed by image analysis of these photomicrographs. Thus, each datum that will be reported here as a typical, illustrative example of a characteristic feature was actually based on an average of measurements of the corresponding feature from at least 10 SEM or FESEM images taken at each of the five randomly picked up locations, as mentioned above. The coating had a nearly uniform thickness of about ~210 ± 6.3 μm (Figure 4.12a). It had a highly heterogeneous, porous microstructure (Figure 4.12b–f). Akin to what has been observed as well by many other researchers [31, 33, 36], the coating was almost infested with varieties of macro- and microcracks, including both inter- and intrasplat cracks (Figure 4.12b–f). It comprised ellipsoidal, well-flattened splats (Figure 4.12b) wherein the intermicropore distance could vary over a range as small as about 1 μm (e.g., AB) to as large as about 20 μm (e.g., CD) (Figure 4.12c). A typical micropore was about 5 μm in diameter (Figure 4.12d), while the size range of microcracks could span a range as wide as 1.5 to 5.5 μm. Further, the presence of the micropore (Figure 4.12d) was typically associated with the presence of other characteristic features, e.g., microcracks or intersplat boundary of 1–5 μm sizes. Nonetheless, the coating had a nearly continuous interface with the substrate (Figure 4.12e) with characteristic planar defects, e.g., pores and cracks present in the cross section (Figure 4.12f), by an amount that was much smaller than the amount of their presence on the plan section. Moreover, based on the data obtained from image analysis, the average volume percent

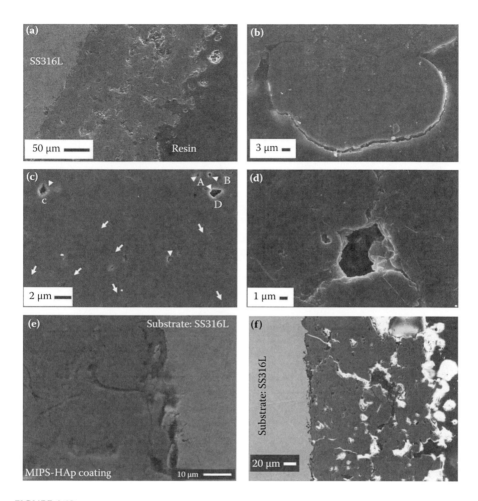

FIGURE 4.12
SEM photomicrographs of the cross section of a polished MIPS-HAp coating: (a) low-magnification view of the coating along with substrate, (b) high-magnification view of well-flattened single splat, (c) typical distribution of microcracks and micropores at high magnification, (d) details of a typical micropore at a higher magnification, (e) details of an interfacial zone, and (f) high-resolution FESEM photomicrograph of the cross section of the MIPS-HAp coating.

open porosity was calculated as ~11% with a range as wide as 4.7–15.6%. Thus, the average volume percent open porosity across the cross section was evidently about half of that measured for the plan section. In addition, the equivalent average spherical diameter of micropores was in a range of 0.9–4.4 µm (Figure 4.13a). Similarly, the average microcrack size was in a range of 2.2–6.6 µm (Figure 4.13b). Further, per unit area, e.g., µm², the number of micropores (e.g., ~2 × 10⁻³) was similar to that (e.g., ~3 × 10⁻³) of microcracks.

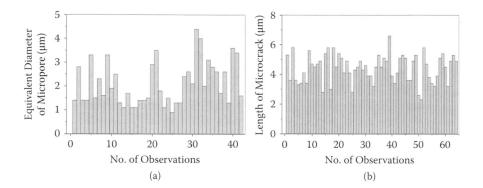

FIGURE 4.13
Distribution of the (a) equivalent diameter of micropores and (d) length of microcracks as a function of the number of observations in the MIPS-HAp coating.

On the other hand, the equivalent average spherical diameter of macropores was in a range as wide as, e.g., 10–75 μm.

4.6 Porosity Dependencies of Young's Modulus and Hardness

Plasma sprayed coatings, e.g., MAPS- or MIPS-HAp coatings, will always have pores. So, the solid load-bearing area will be relatively much less than its bulk counterparts. The question that automatically comes in front is: How will the porosity, i.e., volume fraction open porosity (p), affect the Young's modulus or hardness of such coatings? Based on measurement of Young's modulus by the resonant ultrasound spectroscopy technique [37] for sintered HAp ceramics with ($0.05 < p < 0.51$), it has been found that the conventional exponential relationship [38] is as efficient as the well-known linear equation of porosity dependence of Young's modulus (E):

$$E = E_0 \exp(-bp) \tag{4.2}$$

Further, the rate of decrease in Young's modulus with increasing porosity (dE/dp) was dominated by only the total volume fraction porosity and was insensitive to whether the pore size distribution was unimodal or bimodal [37]. Based on data from our own research and literature data [12, 18, 39–54] for HAp (excluding those from [37]), our effort of using Equation (4.2) to extend the porosity range further (e.g., $0 < p < 0.54$) gave (Figure 4.14a)

$$E = 117.4 \exp(-3.9p) \text{ GPa} \tag{4.3}$$

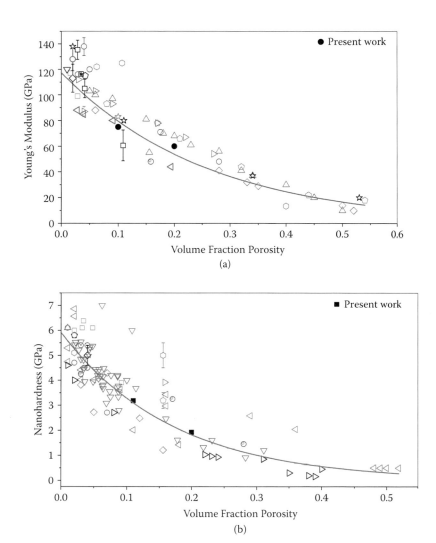

FIGURE 4.14
The porosity dependence of mechanical properties: (a) Young's modulus (data taken from references 12, 18, 39–54) and (b) hardness (data taken from references 12, 39–45, 47, 55–65). The data from the literature as well as the present work as a function of volume fraction open porosity are included. The solid lines in Figure 4.14a and b, respectively, represent the dependence of Young's modulus and hardness on volume fraction porosity according to Equations (4.3) and (4.5).

Here, the correlation coefficient (i.e., r^2) was 0.7. The goodness of fit is not too bad, and in fact could be claimed as reasonably good given the heterogeneity itself present in the sources of this large variety of data [12, 18, 39–54]. Thus, from Equation (4.3), $E_0 = 117$ GPa and $b = 3.9$, which compared favorably with $E_0 = 125$ GPa and b = 3.4 reported in the literature [37]. It is needless to mention that Young's moduli of a material with zero porosity and p are E_0 and E, respectively. Although, classically speaking, E_0 is nothing but the Voigt-Reuss-Hill average of the single crystal elastic constants [48] for a polycrystalline material with randomly oriented grains like the present HAp ceramics being considered [12, 18, 39–54]; it provides actually an "aggregate" value that is an appropriate estimate of the effective Young's modulus of the material. The success of Equation (4.3) also raises another question: Will the nanohardness (H) of sintered polycrystalline HAp ceramics depend on p? It is indeed very fascinating to note that the answer is yes, and the nature of dependency [55] is similar to that of Equation (4.3), i.e.,

$$H = 6 \exp(-6.03p) \text{ GPa } [0.02 < p < 0.31] \tag{4.4}$$

These data were obtained from 42 sintered monophase hydroxyapatite specimens having average grain sizes between 1.7 and 7.4 μm [55]. An attempt to fit the nanohardness data measured for the MIPS-HAp coatings in the present work and a rather huge amount of literature data [12, 39–45, 47, 55–65], excluding those from [55], yielded (Figure 4.14b)

$$H = 5.92 \exp(-5.9p) \text{ GPa } [0.01 < p < 0.52] \tag{4.5}$$

The correlation coefficient (i.e., r^2) improved from 0.7 to 0.83 in this case. Further, from Equation (4.5), $H_0 = 5.92$ GPa and $b = 5.9$, which compared favorably with $H_0 = 6$ GPa and b = 6 reported by other researchers [55]. In addition, based on the Equations (4.3) and (4.5) the brittleness index (E_0/H_0) was ~20, which matched very closely with the value of (E_0/H_0) ~ 21 that can be obtained from relevant data reported in the literature [37, 55]. In addition, the value of ~20 predicted for the brittleness index (E_0/H_0) was also close to the experimentally measured data of ~16 that was obtained from experimentally measured values of Young's modulus (98.2 GPa) and hardness (6.1 GPa). These close matches provided more evidence of efficacy for Equations (4.3) and (4.5) obtained in the present work.

4.7 Qualitative Model for Explanation of Anisotropy

The FESEM photomicrographs for polished plan and cross sections are shown in Figure 4.15. The photomicrograph for a plan section is shown in Figure 4.15a. The photomicrograph for a cross section is shown in Figure 4.15b. The densities of pores, cracks, and defects were much more on the plan section (Figure 4.15a) than on the cross section (Figure 4.15b). Thus, the difference in microstructure between plan and cross sections of a given plasma sprayed coating appeared to be characteristic, and hence deserve

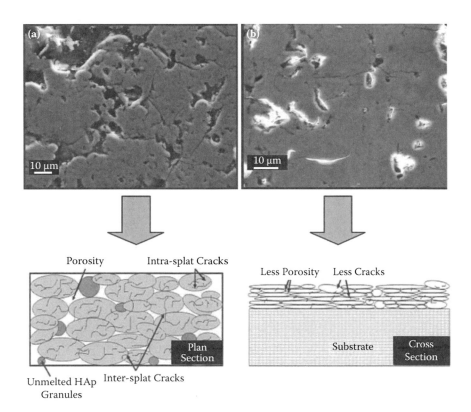

FIGURE 4.15
Proposed logic of surface morphology: FESEM image of the MIPS-HAp coating: (a) plan section along with a schematic of the deposited microstructure and (b) cross section along with a schematic of the deposited microstructure.

a discussion. To take up this issue, we present a schematic of the relevant microstructures in the bottom half of Figure 4.15.

Based on the above evidence, we suggest, as schematically depicted in Figure 4.15a (lower part) for the plan section and in Figure 4.15b (lower part) for the cross section, that the presence of such pores and cracks would certainly reduce the total solid load-bearing contact area [66, 67]. The more the reduction, the more was the likelihood of reduction in nanohardness. As a result, nanohardness and Young's modulus would be lower on the plan section than on the cross section. That is why, at a given load of nanoindentation, the measured value of Young's modulus, as well as hardness, was almost always higher for the cross section and lower for the plan section, and this phenomenon, linked to the relative extent of reduction in solid load-bearing contact area, gave rise to the anisotropy in Young's modulus and hardness in the MIPS-HAp coatings of the present work. In addition, it should be borne in mind that differences in the overall average volume fraction open porosity in plan and cross sections of the coating will also have a direct bearing on the measured nanohardness or Young's modulus data.

4.8 Origin of Modeling on Pore Shape

Several researchers [68–73] had opined that Young's modulus of thermal spray deposits was influenced not only by the volume percent open porosity, but also by the pore morphology. The shape of pore or void could be spherical, elliptical, or a superimposition of spheroidal and ellipsoidal pores or voids altering the aspect ratio. Experimental evidence obtained during the course of the present work had also suggested that the pores could be spherical, elliptical, or penny or thin crack shaped in nature [18]. Therefore, taking this aspect of the microstructure into account, the Young's moduli along the plan section (E_{11}) and across the cross section (E_{22}) of the coating had been predicted by the spherical pore model, elliptical pore model, penny-shaped pore model, and thin crack-shaped pore model.

4.8.1 Modeling of Elastic Constants

Among several mechanical properties, elastic properties such as Young's modulus, shear modulus, bulk modulus, and Poisson's ratio play a pivotal role because a wide range of mechanical properties are related to them. However, Young's moduli of thermal spray deposits are known to be much lower than those of the bulk materials due to their characteristically heterogeneous microstructure. The Young's moduli of MIPS-HAp coating was microstructure dependent, which will be discussed in Chapter 7. More specifically, it depends on the void aspect ratio, volume fraction of void, volume fraction of interlamellar void, volume fraction of interlamellar crack, crack density parameter, etc. Further, the plasma spray deposits exhibit highly anisotropic behavior that originates from the lamellar microstructure. Experimental evidence of such anisotropic behavior was already discussed in Section 4.7. Therefore, it was necessary to consider the measurement directions when determining the Young's modulus of thermally sprayed deposits.

However, elastic constants of solid materials are fundamental and important material properties. Elastic constants include Young's modulus, shear modulus, bulk modulus, and Poisson's ratio, which all are directly related to interatomic bonding. Effective elastic constants are of particular interest when dealing with porous (or cracked) materials and composites, since these microstructural factors are incorporated into the experimentally estimated elastic constants. Thermally sprayed deposits are considered transversely isotropic with respect to the spray direction. The five independent components of the stiffness tensor are required to describe the transversely isotropic case [68–70].

4.8.2 Physical Background of Modeling

In reality, elastic constants of the coating depend on its microstructural factors, such as void or pore shape, void aspect ratio, crack density parameter, etc. The presence of typical spherical and elliptical pores with the corresponding schematics is shown in Figures 4.16a, b and 4.17a, b, respectively. Those voids could be aligned in a unidirectional or two-dimensional fashion in a random manner. Thus, a unidirectionally aligned and two-dimensional randomly oriented void model had been utilized by other researchers to analyze the elastic behavior of porous coatings [68–70]. Similarly, the presence of typical penny-shaped and thin crack-shaped pores with the corresponding schematics are shown in Figures 4.18a, b and 4.19a, b, respectively. Thus, penny-shaped and thin crack-shaped void models had also been utilized by other researchers [68–70] to describe the elastic behavior of porous coatings.

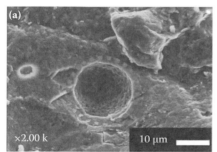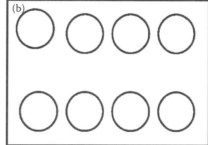

FIGURE 4.16
Presence of typically spherical pore in the MIPS-HAp coating: (a) SEM image (dotted line in SEM image indicates the spherical pores) and (b) schematic of an array of spherical pores.

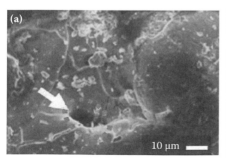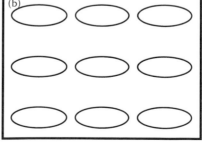

FIGURE 4.17
Presence of typically elliptical pore in the MIPS-HAp coating: (a) SEM image (dotted line in SEM image indicates the elliptical pores) and (b) schematic of elliptical pores.

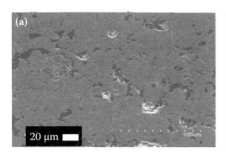
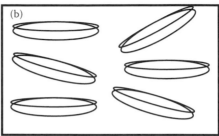

FIGURE 4.18
Presence of typically penny-shaped pores of random orientation in the MIPS-HAp coating:
(a) SEM image (dotted line in SEM image indicates the penny-shaped pores) and (b) schematic
of penny-shaped pores.

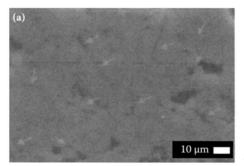
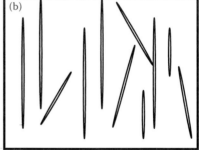

FIGURE 4.19
Presence of typically thin crack-shaped pores of random orientation in the MIPS-HAp coating:
(a) SEM image (arrow in SEM image indicates the crack-shaped pores) and (b) schematic of
crack-shaped pores.

Finally, the five independent elastic constants, e.g., E_{11}, E_{22}, μ_{12}, μ_{23}, and υ, had been modeled following [68–70]. However, we will concentrate here only on elastic modulus data. The models utilized in the present work [18] for MIPS-HAp are:

1. Unidirectionally aligned void model
2. Two-dimensional randomly oriented void model for elliptical and spherical shape of voids
3. Superimposition of both elliptical and spherical voids modeled by penny-shaped void model and thin crack-shaped model

In fact, the voids or pores in the MIPS-HAp coating were neither obviously spherical nor elliptical. Rather, the ultimate microstructure of the MIPS-HAp coating always contained a superposition of both elliptical and spherical voids.

4.8.3 Experimental Validation of the Void Models: Superimposition of Spherical and Elliptical Voids

The data presented in Figure 4.20a show the values of E_{11} as a function of the void volume fraction from 0 to 1 for parametric variation of the aspect ratio values from $a = 0.01$ to $a = 1$. It is clearly evident that, for a given void volume fraction, the value of E_{11} decreased with a decrease in the aspect

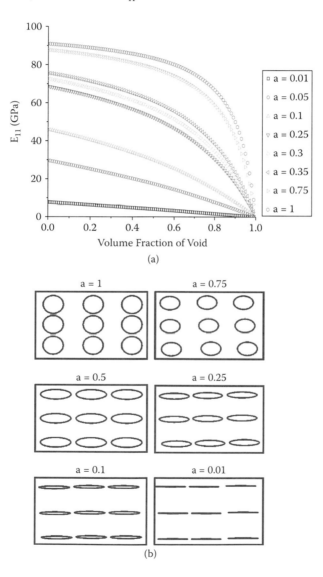

(a)

(b)

FIGURE 4.20 (See color insert.)
(a) Predicated variation of E_{11} as a function of the volume fraction of void. (b) Schematic of variations in pore shapes: void aspect ratio from 0.01 to 1.

ratio. Figure 4.20b presents the corresponding schematic of variations in pore shapes, which depicts the transition from circular (e.g., the case of a spheroidal void in 3D with $a = 1$) to thin crack-like (e.g., the case of a slit void with $a = 0.01$) voids. The normalized value of E_{22} was not dependent on a, and hence it was expected that the normalized value of E_{22} would not vary with variations in the values of a. Further, it was illustrated [18] that values of E_{11}, E_{22} and the anisotropy factor, i.e., the ratio between E_{22} and E_{11} ($(E_{22}/E_{11})_{pred}$ – 1.4), predicted with the penny-shaped pore model matched the best among all models with the experimentally measured data ($(E_{11}/E_{22})_{exp}$ – 1.3), which means the ratio of the modulus data measured in the cross section and plan sections of the present MIPS-HAp coating [18]. Having seen that the coating had adequate nanohardness, Young's modulus, and that the penny-shaped void model would suffice to describe the anisotropy of Young's modulus of the coating, it was felt necessary to investigate what is the intrinsic resistance, i.e., the fracture toughness of the coating, that would be available to act against catastrophic crack propagation at the scale of the local microstructure.

4.9 Summary

In this chapter, we have discussed the porosity, crystallinity, stoichiometry, etc., of HAp coatings. Although both MAPS- and MIPS-HAp coatings have been discussed on a comparative scale, relatively more detailed discussion has been devoted to the latter type of coatings, rather than to the former. The MIPS-HAp coating was nearly phase pure with 80–92% crystallinity and a relatively higher porosity of ~20% in the plan and ~11% across the cross sections. These values of crystallinity and porosity of the MIPS coatings were much higher than those usually reported for the conventional MAPS-HAp coatings. Post-heat treatment improved the degree of crystallization in the coating. The FTIR data indicated that a process of dehydroxylation and rehydroxylation might have occurred in the as-sprayed and post-heat-treated coatings, respectively. SEM and FESEM studies of the coating microstructure revealed the characteristic presence of macro- and microcracks, inter- and intrasplat cracks, macro- and micropores, and unmelted HAp particles. The image analysis technique provided information on various important microstructural parameters, e.g., average splat size, splat aspect ratio, and micropore and microcrack sizes. The coating had a highly heterogeneous microstructure with a splat size of about 50–70 µm, macropore size of about 10–50 µm, and micropore size of about 1 µm. The thickness of the MIPS-HAp coating was measured as ~210 µm. As revealed by FESEM photomicrographs and volume percent porosity data obtained from image analysis of the heterogeneous microstructure, this anisotropy in the data was linked to the larger volume percent porosity, as well as higher spatial density of planar defects, pores, and cracks in the plan section over those in the cross section. In this connection, a qualitative model was schematically developed to pictorially depict the genesis of anisotropy in nanohardness of the present MIPS-HAp coating. Finally, the best fit for the porosity dependence of combined hardness data from literature and the present work was given by an empirical generic equation of the form $X = X_0 \exp(-bp)$, where X stands for nanohardness (H) or Young's modulus (E), as the case may be. The exponential dependencies of E and H on porosity estimated E_0 as 117.4 GPa and H_0 as 5.92 GPa, which were comparable to literature data, where the suffix 0 stands for theoretically dense, e.g., zero-porosity material. Thus, the volume fraction open porosity also played an important role in anisotropy of nanohardness in the present MIPS-HAp coating. Young's modulus of thermal spray deposits is influenced not only by the volume percent of open porosity, but also by the pore morphology. The shape of pore or void could be spherical, elliptical, or a superimposition of spherical and elliptical pores/voids altering the aspect ratio. Experimental evidence gathered for the present MIPS-HAp coating also suggested that the pores could be spherical, elliptical, or penny or thin crack shaped in nature. Therefore, taking this aspect of the microstructure into account, the Young's moduli along the plan section (E_{11}) and cross section (E_{22}) of the coating had been predicted by the spherical pore model, elliptical pore model, penny-shaped pore model, and the thin crack-shaped pore model. Finally, it was illustrated that values of E_{11}, E_{22} and the anisotropy factor, i.e., the ratio between E_{22} and E_{11}, predicted with the penny-shaped pore model matched the best among all models with the experimentally measured data of the present MIPS-HAp coatings.

In vitro studies of HAp coatings, e.g., in particular the investigations of HAp coatings in accelerated simulated body fluid (SBF) medium, will be discussed in Chapter 5. The experimental evaluations of the phase analysis, spectroscopic observation and microstructural evolution of HAp coatings in general and MIPS-HAp coatings in particular, will be studied both before and after SBF immersion. The context of dissolution of Ca and P shall be elaborated. The possibility of leaching out for metal ions from the biomedical implants immersed in SBF shall also be critically examined in Chapter 5.

References

1. Yang Y. C. and Chang E. 2001. Influence of residual stress on bonding strength and fracture of plasma-sprayed hydroxyapatite coatings on Ti-6Al-4V substrate. *Biomaterials*, 22: 1827–1836.
2. Yang Y. C. and Chang E. 2003. The bonding of plasma-sprayed hydroxyapatite coatings to titanium: effect of processing, porosity and residual stress. *Thin Solid Films*, 444: 260–275.

3. Yang Y. C. 2007. Influence of residual stress on bonding strength of the plasma-sprayed hydroxyapatite coating after the vacuum heat treatment. *Surface and Coatings Technology*, 201: 7187–7193.

4. Willmann G. 1999. Coating of implants with hydroxyapatite material connections between bone and metal. *Advanced Engineering Materials*, 1: 95–105.

5. Yang C. Y., Lin R. M., Wang B. C., Lee T. M., Chang E., Hang Y. S., and Chen P. Q. 1997. In vitro and in vivo mechanical evaluations of plasma-sprayed hydroxyapatite coatings on titanium implants: the effect of coating characteristics. *Journal of Biomedical Materials Research*, 37: 335–345.

6. Yang C. Y., Wang C., Chang E., and Wu J. D. 1995. Bond degradation at the plasma-sprayed HA coating/Ti-6Al-4V/alloy interface: an in vitro study. *Journal of Materials Science: Materials in Medicine*, 6: 258–265.

7. Wang B. C., Chang E., Yang C. Y., Tu D., and Tasi C. H. 1993. Characteristics and osteoconduction of three different plasma-sprayed hydroxyapatite-coated titanium implants. *Surface and Coatings Technology*, 58: 107–117.

8. Zheng X., Huang M., and Ding C. 2000. Bond strength of plasma-sprayed hydroxyapatite/Ti composite coatings. *Biomaterials*, 21: 841–849.

9. Ding S. J., Su Y. M., Ju C. P., and Lin J. H. C. 2001. Structure and immersion behavior of plasma-sprayed apatite-matrix coatings. *Biomaterials*, 22: 833–845.

10. Chen C. C., Huang T. H., Kao C. T., and Ding S. J. 2004. Electrochemical study of the in vitro degradation of plasma-sprayed hydroxyapatite/bioactive glass composite coatings after heat treatment. *Electrochimca Acta*, 50: 1023–1029.

11. Yang C. Y., Chen C. R., Chang E., and Lee T. M. 2007. Characteristics of hydroxyapatite coated titanium porous coatings on Ti-6Al-4V substrates by plasma sprayed method. *Journal of Biomedical Materials Research Part B*, 82: 450–459.

12. Gu Y. W., Khor K. A., and Cheang P. 2003. In vitro studies of plasma-sprayed hydroxyapatite/Ti-6Al-4V composite coatings in simulated body fluid (SBF). *Biomaterials*, 24: 1603–1611.

13. Balamurugan A., Balossier G., Kannan S., Michel J., Faure J., and Rajeswari S. 2007. Electrochemical and structural characterisation of zirconia reinforced hydroxyapatite bioceramic sol-gel coatings on surgical grade 316L SS for biomedical applications. *Ceramics International*, 33: 605–614.

14. Weng W. and Baptista J. L. 1999. Preparation and characterization of hydroxyapatite coatings on Ti6Al4V alloy by a sol-gel method. *Journal of the American Ceramic Society*, 82: 27–32.

15. Manso-Silvan M., Langlet M., Jimenez C., Fernandez M., and Martinez-Duart J. M. 2003. Calcium phosphate coatings prepared by aerosol-gel. *Journal of the European Ceramic Society*, 23: 243–246.

16. Cheng K., Zhang S., Weng W., Khor K. A., Miao S., and Wang Y. 2008. The adhesion strength and residual stress of colloidal-sol gel derived [beta]-tricalcium-phosphate/fluoridated-hydroxyapatite biphasic coatings. *Thin Solid Films*, 516: 3251–3255.

17. Kim H. M., Miyaji F., Kokubo T., and Nakamura T. 1997. Bonding strength of bonelike apatite layer to Ti metal substrate. *Journal of Biomedical Materials Research Part B*, 38: 121–127.

18. Dey A. 2011. Physico-chemical and mechanical characterization of bioactive ceramic coating. PhD dissertation, Indian Institute and Engineering Technology (formerly Bengal Engineering and Science University), Shibpur, Howrah, India.

19. Dey A., Mukhopadhyay A. K., Gangadharan S., Sinha M. K., and Basu D. 2009. Characterization of microplasma sprayed hydroxyapatite coating. *Journal of Thermal Spray Technology*, 18: 578–592.
20. Dey A., Mukhopadhyay A. K., Gangadharan S., Sinha M. K., and Basu D. 2009. Development of hydroxyapatite coating by microplasma spraying. *Materials and Manufacturing Processes*, 24: 1321–1330.
21. Landi E., Tampieri A., Celotti G., and Sprio S. 2000. Densification behaviour and mechanisms of synthetic hydroxyapatites. *Journal of the European Ceramic Society*, 20: 2377–2387.
22. Han Y., Li S., Wang X., and Chen X. 2004. Synthesis and sintering of nanocrystalline hydroxyapatite powders by citric acid sol-gel combustion method. *Materials Research Bulletin*, 39: 25–32.
23. Sridhar T. M., Mudali U. K., and Subbaiyan M. 2003. Sintering atmosphere and temperature effects on hydroxyapatite coated type 316L stainless steel. *Corrosion Science*, 45: 2337–2359.
24. Chen C. C. and Ding S. J. 2006. Effect of heat treatment on characteristics of plasma sprayed hydroxyapatite coating. *Materials Transactions*, 47: 935–940.
25. Sun L., Berndt C. C., and Grey C. P. 2003. Phase structural investigations of plasma sprayed hydroxyapatite coating. *Materials Science and Engineering A*, 360: 70–84.
26. Ding S. J., Hung T. H., and Kao C. T. 2003. Immersion behaviour of plasma sprayed modified hydroxyapatite coatings after heat treatment. *Surface and Coating Technology*, 165: 248–257.
27. Nath S., Biswas K., and Basu B. 2008. Phase stability and microstructure development in hydroxyapatite-mullite system. *Scripta Materialia*, 58: 1054–1057.
28. Mahabole M. P., Aiyer R. C., Ramakrishna C. V., Sreedhar B., and Khairnar R. S. 2005. Synthesis, characterization and gas sensing property of hydroxyapatite ceramic. *Bulletin of Materials Science*, 28: 535–545.
29. Morales J. G., Burgues J. T., Boix T., Fraile J., and Clemente R. R. 2001. Precipitation of stoichiometric hydroxyapatite by a continuous method. *Crystal Research and Technology*, 36: 15–26.
30. Hench L. L. and Wilson J. 1993. *An introduction to bioceramics: advance series in ceramics*, vol. 1. Singapore: World Scientific Publishing.
31. Kweh S. W. K., Khor K. A., and Cheang P. 2000. Plasma-sprayed hydroxyapatite (HA) coatings with flame-spheroidized feedstock: microstructure and mechanical properties. *Biomaterials*, 21: 1223–1234.
32. Li H., Khor K. A., and Cheang P. 2002. Titanium dioxide reinforced hydroxyapatite coatings deposited by high velocity oxy-fuel (HVOF) spray. *Biomaterials*, 23: 85–91.
33. Wang M., Yang X. Y., Khor K. A., and Wang Y. 1999. Preparation and characterization of bioactive monolayer and functionally graded coatings. *Journal of Materials Science: Materials in Medicine*, 10: 269–273.
34. Dey A., Mukhopadhyay A. K., Gangadharan S., Sinha M. K., and Basu D. 2009. Weibull modulus of nano-hardness and elastic modulus of hydroxyapatite coating. *Journal of Materials Science*, 44: 4911–4918.
35. Heimann R. B. 2006. Thermal spraying of biomaterials. *Surface and Coatings Technology*, 201: 2012–2019.

36. Li H., Khor K. A., and Cheang P. 2002. Young's modulus and fracture toughness determination of high velocity oxy-fuel-sprayed bioceramic coatings. *Surface and Coatings Technology*, 155: 21–32.
37. Ren F., Case E. D., Morrison A., Tafesse M., and Baumann M. J. 2009. Resonant ultrasound spectroscopy measurement of Young's modulus, shear modulus and Poisson's ratio as a function of porosity for alumina and hydroxyapatite. *Philosophical Magazine*, 89: 1163–1182.
38. Spriggs R. M. 1961. Expression for effect of porosity on elastic modulus of polycrystalline refractory materials, particularly aluminum oxide. *Journal of the American Ceramic Society*, 44: 628–629.
39. Kumar R. R. and Wang M. 2002. Modulus and hardness evaluations of sintered bioceramic powders and functionally graded bioactive composites by nano-indentation technique. *Materials Science and Engineering A*, 338: 230–236.
40. Kundu B., Mukhopadyay A. K., Dey A., Gangadharan S., and Basu D. 2006. Depth sensitive indentation behaviour of sintered HAP compacts. In *Proceedings on Biomaterials and Biomedical Devices (BMD-06)*, Kolkata, India, December 12–13.
41. Nath S., Dey A., Mukhopadhyay A. K., and Basu B. 2009. Nanoindentation response of novel hydroxyapatite-mullite composites. *Materials Science and Engineering A*, 513–514: 197–201.
42. Mukhopadyay A. K., Seal S., Sinha M. K., Kundu B., Dey A., Gangadharan S., and Basu D. 2006. Micromechanical characterization of plasma sprayed HAP coating. In *Proceedings on Biomaterials and Biomedical Devices (BMD-06)*, Kolkata, India, December 12–13.
43. Kumar R., Cheang P., and Khor K. A. 2003. Spark plasma sintering and in vitro study of ultra-fine HA and ZrO_2-HA powders. *Journal of Materials Processing Technology*, 140: 420–425.
44. Kim H. W., Knowles J. C., Li L. H., and Kim H. E. 2005. Mechanical performance and osteoblast-like cell responses of fluorine-substituted hydroxyapatite and zirconia dense composite. *Journal of Biomedical Materials Research Part A*, 72: 258–268.
45. Asmus S. M. F., Sakakura S., and Pezzotti G. 2003. Hydroxyapatite toughened by silver inclusions. *Journal of Composite Materials*, 37: 2117–2129.
46. Tang C. Y., Uskokovic P. S., Tsui C. P., Veljovic D., Petrovic R., and Janackovic D. 2009. Influence of microstructure and phase composition on the nanoindentation characterization of bioceramic materials based on hydroxyapatite. *Ceramics International*, 35: 2171–2178.
47. Chen B., Zhang T., Zhang J., Lin Q., and Jiang D. 2008. Microstructure and mechanical properties of hydroxyapatite obtained by gel-casting process. *Ceramics International*, 34: 359–364.
48. Landolt H. and Bornstein R. 1979. *Elektrische, piezoelektrische, pyroelektrische, piezooptische, elektrooptische konstanten und nichtlineare dielektrische suszeptibilitaten, zahlenwerte und funktionen aus naturwissenschaften und technik*, Gruppe III, Bd. 11. Berlin: Springer-Verlag.
49. Akao M., Aoki H., and Kato K. 1981. Mechanical properties of sintered hydroxyapatite for prosthetic applications. *Journal of Materials Science*, 16: 809–812.
50. With G. D., Dijk H. V., Hattu N., and Prijs K. 1981. Preparation, microstructure and mechanical properties of dense polycrystalline hydroxyapatite. *Journal of Materials Science*, 16: 1592–1598.

51. Arita I., Wilkinson D., Mondragon M., and Castano V. 1995. Chemistry and sintering behaviour of thin hydroxyapatite ceramics with controlled porosity. *Biomaterials*, 16: 403–408.
52. Liu M. D. 1998. Preparation and characterisation of porous hydroxyapatite bio-ceramic via a slip-casting route. *Ceramics International*, 24: 441–446.
53. Charriere E., Terrazzoni S., Pittet C., Mordasini P., Dutoit M., Lemaitre J., and Zysset P. 2001. Mechanical characterization of brushite and hydroxyapatite cements. *Biomaterials*, 22: 2937–2945.
54. He L. H., Standard O. C., Huang T. T. Y., Latella B. A., and Swain M. V. 2008. Mechanical behaviour of porous hydroxyapatite. *Acta Biomaterialia*, 4: 577–586.
55. Hoepfner T. P. and Case E. D. 2003. The influence of the microstructure on the hardness of sintered hydroxyapatite. *Ceramics International*, 29: 699–706.
56. Mancini C. E., Berndt C. C, Sun L., and Kucuk A. 2001. Porosity determinations in thermally sprayed hydroxyapatite coatings. *Journal of Materials Science*, 36: 3891–3896.
57. Nieh T. G., Choi B. W., and Jankowski A. F. 2001. Synthesis and characterization of porous hydroxyapatite and hydroxyapatite coatings. Report submitted to Minerals, Metals and Materials Society Annual Meeting and Exhibition, Los Angeles.
58. Dey A. and Mukhopadhyay A. K. 2010. Anisotropy in nano-hardness of micro-plasma sprayed hydroxyapatite coating. *Advances in Applied Ceramics*, 109: 346–354.
59. Khalil K. A., Kim H. Y., Kim S. W., and Kim K. W. 2007. Observation of tough-ness improvement of the hydroxyapatite bioceramics densified using high-frequency induction heat sintering. *International Journal of Applied Ceramic Technology*, 4: 30–37.
60. Ramesh S., Tan C. Y., Bhaduri S. B., and Teng W. D. 2007. Rapid densifica-tion of nanocrystalline hydroxyapatite for biomedical applications. *Ceramics International*, 33: 1363–1367.
61. Thangamani N., Chinnakali K., and Gnanam F. D. 2002. The effect of pow-der processing on densification, microstructure and mechanical properties of hydroxyapatite. *Ceramics International*, 28: 355–362.
62. Muralithran G. and Ramesh S. 2000. The effects of sintering temperature on the properties of hydroxyapatite. *Ceramics International*, 26: 221–230.
63. Best S. and Bonfield W. 1994. Processing behaviour of hydroxyapatite powders with contrasting morphology. *Journal of Materials Science: Materials in Medicine*, 5: 516–521.
64. Wang P. E. and Chaki T. K. 1993. Sintering behaviour and mechanical prop-erties of hydroxyapatite and dicalcium phosphate. *Journal of Materials Science: Materials in Medicine*, 4: 150–158.
65. Slosarczyk A., Stobierska E., Paskiewicz Z., and Gawlicki M. 1996. Calcium phosphate materials prepared from precipitates with various calcium: phos-phorus molar ratios. *Journal of the American Ceramic Society*, 79: 2539–2544.
66. Mukhopadhyay A. K. and Phani K. K. 1998. Young's modulus-porosity rela-tions—an analysis based on minimum contact area model. *Journal of Materials Science*, 33: 69–72.
67. Rossi R. C. 1968. Prediction of the elastic moduli of composites. *Journal of the American Ceramic Society*, 51: 433–439.

68. Leigh S. H. and Berndt C. C. 1999. Quantitative evaluation of void distributions within a plasma-sprayed ceramic. *Journal of the American Ceramic Society*, 82: 17–21.
69. Leigh S. H. and Berndt C. C. 1999. Modelling of elastic constants of plasma spray deposits with ellipsoid-shaped voids. *Acta Materialia*, 47: 1575–1586.
70. Leigh S. H., Lee G. C., and Berndt C. C. 1998. Modelling of elastic constants of plasma spray deposits with spheroid shaped voids. In *Thermal spray meeting: the challenges of the 21st century*, Coddet C. (Ed.). Materials Park, OH: ASM International, 587–592.
71. Parthasarathi S., Tittmann B. R., Sampath K., and Onesto E. J. 1995. Ultrasonic characterization of elastic anisotropy in plasma-sprayed alumina coatings. *Journal of Thermal Spray Technology*, 4: 367–373.
72. Tobe S., Kodama S., Misawa H., and Ishikawa K. 1991. Rolling fatigue behavior of plasma sprayed coatings on aluminum alloy. In *Thermal spray research and application*, Beruecki T. E. (Ed.). Materials Park, OH: ASM International, 171–177.
73. Lauschmann H., Moravcova M., Neufuss K., and Chraska P. 1994. Elastic Young's modulus of plasma sprayed materials. In *Thermal spray industrial applications*, Berndt C. C. and Sampath S. (Eds.). Materials Park, OH: ASM International, 699–701.

5

In Vitro Studies of Hydroxyapatite Coatings

5.1 Introduction

This chapter is about the in vitro response of hydroxyapatite (HAp) coatings, in particular in a simulated body fluid (SBF) environment. Before any in vivo trial, the efficacy of the bioactive coating should be verified in an accelerated environment.

In vitro tests can be performed in a cultured cell medium or a salt-containing solution. The first type of test assesses biocompatibility by observing the behavior of cells in the presence of the material. The types of response that indicate toxicity are cell death, reduced cell adhesion, altered cell morphology, reduced cell proliferation, and reduced biosynthetic activity [1]. Salt solutions that attempt to duplicate body conditions are used to observe the behavior of biomaterials after certain immersion periods. The second type of test is solely used to study material changes. Biological implants require a surface that is compatible with the body environment. Plasma spraying is a well-known technique that has been chosen to produce coatings varying from about 50 to 400 μm thickness for biological applications [2]. The process is clean, and the high rates of deposition allow coatings to be produced fairly rapidly. Since thermal spraying is a direct line-of-sight process, isolated areas on an implant can be coated, or the entire surface of a complex geometry can be coated by rotating the object. Any substrate can be used, but for reasons of practicality, ceramics or metals are chosen. The process enables control over properties such as porosity, surface morphology, roughness, composition, and crystallinity, which in turn influence the chemical, physical, and mechanical properties.

The concept of the SBF dissolution investigation dates back to ancient history. Even in the old days, during a battle a bone fracture was commonplace for soldiers. It also happened to common men after an accident. If and when these fractures happened, both soldiers and common people used to wind up their broken place with a wooden stick or plate. When they would visit a doctor, he would almost obviously prescribe complete bed rest. The most interesting part of the story is that after some weeks, the fracture of the bone was found to have healed almost automatically, without any surgery or any external medication. People initially used to believe this was a magical performance of the concerned physician. Now, the question that genuinely springs up in our mind is: What is the reason for "self-healing" or "magic"? Since we belong to a scientific era, we cannot accept magic as the answer. So, we need to know the scientific reason behind such self-healing capacity of the human body. Actually, what happened is this that the blood plasma produced inorganic calcium phosphate apatite-based filler material. This filler material healed the fractured part of the bone. This science was clearer to the researchers when they were able to grow calcium phosphate by the biomimetic process, as mentioned in Chapter 2. In a biomimetic process, the simulated body fluid is produced at a temperature of 37°C in the laboratory, and subsequently the substrate is immersed in it, maintaining the pH at 7.4. After a certain duration, the nanometric calcium phosphate (CaP) apatite forms by a heterogeneous nucleation process, as mentioned in Chapter 2, on the surface of the substrate. This background information will undoubtedly help us to understand why the in vitro SBF immersion study is important prior to an animal or human trial of the coating.

5.2 Literature Status

Indeed, the researchers have done a lot of work on SBF immersion studies in conventionally macroplasma sprayed (MAPS) and other CaP and hydroxyapatite coatings developed by other techniques [2–16]. They have reported extensively on the structural changes as a function of post-heat treatment, coating thickness, type of substrates, duration of the immersion, etc. Further, mechanical properties like bonding strength of the coating, hardness, Young's modulus, etc., have been evaluated to understand the efficacy of the coating in an accelerated biological environment. However, SBF immersion studies with microplasma sprayed (MIPS) HAp coatings have rarely been attempted [17–20].

Zhao et al. [17] reported the in vitro investigation of MIPS-coated HAp coatings deposited on Ti-6Al-4V substrates. They had kept the coatings immersed up to 14 days in SBF to evaluate their bioactivity. It was found that a layer of carbonated-apatite covering almost the entire surface of the HAp coatings had indeed developed, and it did not exhibit significant spalling after incubation in SBF. In another study, Zhao et al. [18] conducted a Hanks' balanced salt solution (HBSS) immersion test to evaluate the bioactivity of the MIPS-coated HAp coatings developed on Ti-6Al-4V substrates. A double-layer coating was tried. The first bottom layer was thick and highly crystalline. The layer deposited on top of this was thin, porous, discontinuous, and less crystalline. Such a coating architecture provided the growth of a bone-like apatite layer on top of the HAp coatings on the top. This bone-like apatite layer had formed rapidly during the early stage of immersion. It was suggested that long-term stability of such a double-layer coating was expected to be better due to the presence of the highly crystalline, thick HAp coating at the bottom. In contrast, the results recently published by the present authors, Dey and Mukhopadhyay [19], have demonstrated how the phase-pure and porous MIPS-HAp coatings on SS316L biomedical implant-grade substrates showed excellent response in a SBF environment in terms of nanomechanical properties, e.g., nanohardness and Young's modulus, as well as tribological properties. These results provided experimental proof in support of its biological efficacy prior to the in vivo trial.

5.3 Synthesis of SBF in the Laboratory

SBF was prepared in the present work according to Kokubo formulation [21], which is a buffered (pH of 7.4 at 25°C) solution with an ionic concentration close to that found in the human blood plasma (Table 5.1). Experiments were performed by immersing each HAp-coated sample in 30 ml of SBF solutions using 50 ml capacity polypropylene flasks, thermostatized (37 ± 0.5°C) and statically kept up to different time intervals at that temperature for 1, 4, 7, and 14 days. The SBF dissolution study on the HAp coating was conducted according to the ISO 10993-14 standard [22] by the inductively coupled plasma atomic emission spectroscopy (ICP-AES) method.

TABLE 5.1

Inorganic Composition of Human Blood Plasma (HBP) and the Simulated Body Fluid (SBF) Solution Utilized in the Present Work

Ions	Blood Plasma (ppm)	SBF (ppm)
Na^+	142.0	142.6
Cl^-	103.0	105.0
HCO_3^-	27.0	25.6
K^+	5.0	5.1
Mg^{2+}	1.5	1.5
Ca^{2+}	2.5	2.5
HPO_4^{2-}	1.0	1.0
SO_4^{2-}	0.5	0.5

Source: Reprinted from Dey and Mukhopadhyay, *International Journal of Applied Ceramic Technology*, 11: 65–82, 2013. With permission from American Ceramic Society and Wiley.

5.4 SBF Immersion of MAPS-HAp Coatings on SS316L

If we look into the microstructures presented in Figure 5.1a, b, we can understand there are significant changes in the morphology of the MAPS process-coated HAp coatings on biomedical-grade SS316L substrates after the SBF immersion. The bigger cracks have been observed in the entire microstructure, as has been also reported in literature [2] for HAp coatings deposited by the conventional MAPS process. However, the most important observation was not the presence of the cracks, which were expected to happen anyway, but the deposition of spherical globular CaP apatite (Figure 5.1a). Features of a size less than a micron could be amply noticed in the microstructure when viewed at a higher magnification (Figure 5.1b). The presence of the fine-needle-like apatite structure corroborated well with the energy-dispersive X-ray spectroscopy (EDX) data (Figure 5.2), which confirmed the presence of only the CaP compound in the biomimetically deposited layer that had formed during and following the SBF immersion. These experimental observations were similar to the observations reported by other researchers [2–16]. As mentioned earlier, in a very recent effort, Dey and his coworkers [19, 20] developed novel HAp coatings. These novel coatings were deposited on biomedical-grade SS316L substrates by the MIPS technique, rather than the conventional MAPS technique. The idea behind such an effort was to develop bioactive ceramic coatings for futuristic applications. Therefore, in this chapter we shall describe in detail the results of the SBF immersion studies for these MIPS-HAp coatings deposited on the SS316L substrates.

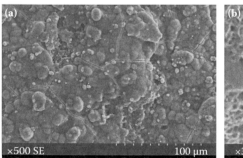 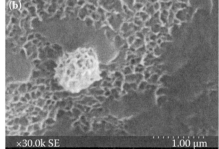

FIGURE 5.1
Microstructures of MAPS-HAp coatings on SS316L after 14 days of immersion in SBF medium: (a) lower and (b) higher magnification shows the change of microstructure and formation of the apatite layer.

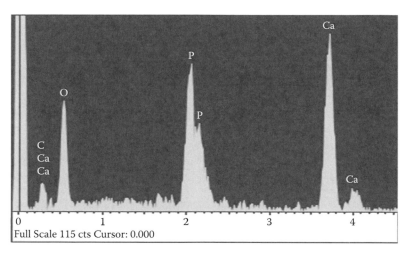

FIGURE 5.2
The EDX spectra of MAPS-HAp coatings on SS316L after 14 days of immersion in SBF medium confirm the presence of CaP compounds.

5.5 SBF Immersion of MIPS-HAp Coatings on SS316L

The results presented in Table 5.2 show that due to formation of an apatite layer after the immersion in the SBF solution over a preselected period of time, e.g., 1, 4, 7, and 14 days, all the MIPS-HAp-coated samples had gained weight (Table 5.2). Now let us look at the experimental results obtained by the ICP-AES technique for the calcium (Ca^{2+}) and potassium (P^{5+}) ion concentrations in the immersion solutions (Figure 5.3a) [19, 20]. The data presented

TABLE 5.2

Weight Enhancement, d Spacing, and Degree of Crystallinity after Immersion in SBF Solution

Type of coatings ➡	As Deposited	No. of Days into SBF Immersion			
		1	4	7	14
Properties ⬇					
Weight enhancement (g)	—	0.0022	0.0019	0.0027	0.0015
d spacing (Å) of (211) plane	2.81046	2.8086	2.80447	2.80659	2.82032
Degree of crystallinity (%)	81.6	71.4	76.5	75.3	61.3
Ca/P ratio	1.67	1.68	1.59	1.57	1.82

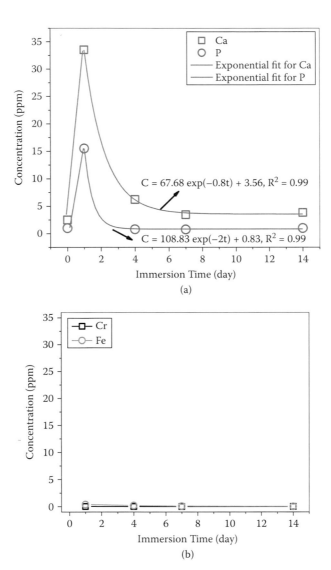

FIGURE 5.3 (See color insert.)
(a) Change of Ca and P concentration with time of immersion shows the dissolution followed by a deposition of CaP compounds. (*C* and *t* stand for concentration and immersion time, respectively.) (Reprinted from Dey and Mukhopadhyay, *International Journal of Applied Ceramic Technology*, 11: 65–82, 2013. With permission from American Ceramic Society and Wiley.) (b) Chromium and iron leaching show almost nil concentration even after 14 days of accelerated dissolution test in a SBF environment.

in Figure 5.3a show that for both calcium and phosphorous ions, only up to the first day following immersion in the SBF solution was the rate of dissolution rapid. It was followed by a sluggish rate of dissolution up to the fourth day. The most interesting part of the observation is that beyond the fourth day, even up to 2 weeks of immersion time, the rate of change in the rate of dissolution with time continued to be ever decreasing. These experimental data confirmed three major physical processes that had happened in the SBF solution. The first was that the dissolution of the MIPS-HAp coating had happened only up to the first day of immersion. The second, and probably the most important fact, is that from the second day onward, up to 2 weeks of immersion time, the continuous apatite deposition process had happened on the surface of the MIPS-HAp coating. The third aspect is that the patterns of decay in calcium and phosphorous ions during the period of the 2nd to 14th days in SBF solution were exponential in nature. Similar exponential patterns were reported for calcium and phosphorus ions when a MAPS-HAp coating was immersed in a SBF solution [3]. It also needs to be checked out whether chromium and iron ions leach out because a possibility definitely exists that they may leach out from the biomedical SS316L-based implant in actual in vivo application. Interestingly, however, the data presented in Figure 5.3b for the MIPS-HAp coatings confirmed that the amount of chromium and iron leaching was almost nil, even after 14 days of accelerated dissolution test in the SBF solution.

Now, let us have a look at the X-ray diffraction (XRD) patterns of the MIPS-HAp coating before and after immersion in SBF solution [19, 20]. These data are shown in Figure 5.4a and 5.4b–e, respectively, for situations before and after the immersion in the SBF solution. The characteristic peaks corresponding to (211), (112), and (300) planes had happened at the corresponding characteristic Bragg's angles of 31.78°, 32.18°, and 32.93°. These data confirmed the purity of the HAp phase. Before immersion in SBF, two additional, yet very minor, peaks of crystalline alpha-tricalcium phosphate (α-TCP) and tetracalcium phosphate (TTCP) were identified. These peaks disappeared after SBF immersion. Similar observation was also reported by other researchers [3] for the dense MAPS-HAp coatings. These impurity phases or non-HAp phases, e.g., TCP and TTCP, are much more soluble than the pure HAp, and these are further dissolved in SBF solution [2, 3]. In the case of conventional MAPS-HAp coatings, the amount of impurity phases like TCP, TTCP, and calcium oxide (CaO) are much more than those usually obtained for the MIPS-HAp coatings. However, we have also seen that even for MIPS-HAp coatings, the peaks corresponding to the extraneous phases, as mentioned above, are removed after post-annealing [19, 20]. It is most interesting to note that as the duration of immersion in the SBF solution increased, so did the prominence of the peak corresponding to the characteristic Bragg angle of 25.80, which corresponds to the presence of a classical carbonated apatite or octacalcium phosphate (OCP) phase. These data confirmed the deposition of the apatite layer. Similar observations have been reported by many other

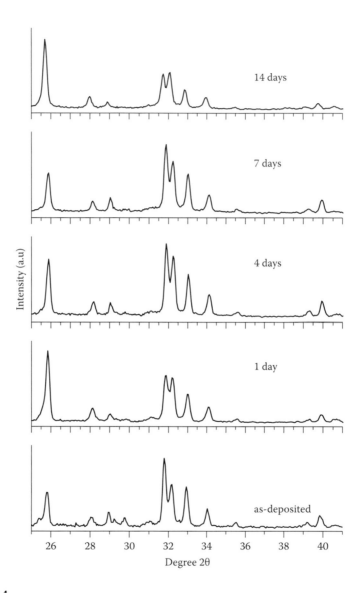

FIGURE 5.4
XRD pattern of the HAp coating on SS316L substrate immersed in SBF solution after (a) 0, (b) 1, (c) 4, (d) 7, and (e) 14 days showed the formation of carbonated apatite or OCP layer after immersion in SBF.

researchers as well [2–4]. It needs to be further emphasized here that the crystallographic structures of carbonated apatite, OCP, and HAp are almost similar [3–5]. Before the immersion, crystallinity of the MIPS-HAp coatings deposited on the SS316L substrates was ~81%. The data on change in crystallinity are presented in Table 5.2 as a function of immersion time in the SBF solution. The same table also presents the data on *d* spacing. For ease of comparison, the *d* values corresponding to the (211) plane were considered, as they characteristically correspond to the maximum (e.g., 100%) intensity. The data presented in Table 5.2 show that the crystallinity of the MIPS-HAp coating had decreased slightly from 71% on day 1 to 61% on day 14 as the time of immersion in the SBF solution was enhanced. The data on changes in the *d* values (Table 5.2) corroborated these data on a decrease in crystallinity with enhancement in immersion time, as mentioned above.

Further, the Fourier transform infrared spectroscopy (FTIR) spectra of the coating before and after immersion in the SBF solution are shown in Figure 5.5a–c [19, 20]. The spectra were comparable with those reported by other researchers [7, 8, 23, 24]. The data presented in Figure 5.5b, c are

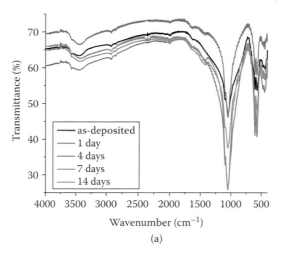

(a)

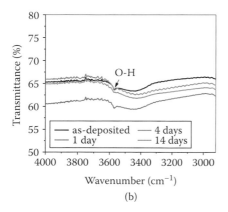

(b)

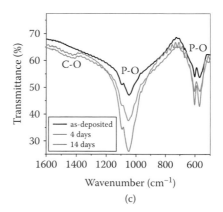

(c)

FIGURE 5.5 (See color insert.)
(a) FTIR spectra of the MIPS-HAp coating before and after SBF immersion. Enlarged view showing the location of (b) OH and (b) CO_3 and PO_4 characteristic peaks. (Reprinted from Dey and Mukhopadhyay, *International Journal of Applied Ceramic Technology*, 11: 65–82, 2013. With permission from American Ceramic Society and Wiley.)

an enlarged view of the data presented in Figure 5.5a. These enlarged views show the characteristic surface active groups, e.g., OH⁻ (at 3571 cm⁻¹, Figure 5.5b) and PO_4^{3-} (at 1092, 1045, and 962 cm⁻¹ for stretching vibration, and 602 and 569 cm⁻¹ for bending vibration, Figure 5.5c), and complete absence of the characteristic peak of CO_3^{2-} for the MIPS-HAp coatings prior to the immersion in the SBF solution. Following immersion in the SBF solution, however, the peak corresponding to the CO_3^{2-} group was noted to exist at 1405–1410 cm⁻¹, as would be expected from the characteristic criterion for the carbonate group. Similar observations were reported by other researchers [7, 8, 23, 24] from FTIR spectra of HAp coatings exposed to immersion in the SBF solutions. The same spectral data, however, also confirmed the presence of the additional characteristic peaks corresponding to the OH⁻ group at 3571–3573 cm⁻¹, for stretching vibrations of the PO_4^{3-} group at 1093–1096 cm⁻¹ and 1048–1050 cm⁻¹, and for bending vibrations of the PO_4^{3-} group at 962, 603, and 569–570 cm⁻¹.

Environmental scanning electron microscopy (E-SEM) photomicrographs of the MIPS-HAp coatings after SBF immersion are shown in Figure 5.6a–c. The main advantages of the E-SEM mode are:

1. It does not require a conducting coating on the sample.
2. There was no charging effect.
3. As there is no coating on the surface of the sample, it reduces the probability of artifacts.

Initially, the coatings appeared with well-defined splats, pores, and cracks. These features are quite common for the thermally sprayed coatings. The surface cracks were comparatively less pronounced after the first day of immersion in SBF solution. After 4 days of immersion in the SBF solution,

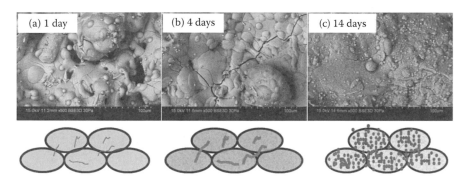

FIGURE 5.6
E-SEM of the MIPS-HAp coatings after immersion in SBF at (a) 1 day, (b) 4 days, and (c) 14 days. Below: Corresponding schematic representation of the changes of the microstructure after immersion of the MIPS-HAp coatings into SBF.

deposition of white microspheres on the surface of the MIPS-HAp coating immersed in SBF solution was also observed. The surface cracks were bigger after 4 days of immersion in the SBF solution. On the other hand, the amount of white microspheres, as well as voids, was increasing as the number of days of immersion increased from the 4th to the 14th day. Similar observation of the disappearance of cracks was also reported by other workers [2]. It was found to have occurred after long duration of immersion in the SBF solution. It was suggested that the CaP compound had deposited as white microspheres [2].

The conventional SEM images of the MIPS-HAp coatings after the first day of immersion in the SBF solution are shown in Figure 5.7a, b. To understand the evolution of microstructure in a SBF environment, the MIPS-HAp-coated sample was placed perpendicularly in the SBF solution. Thus, the photomicrograph presented in Figure 5.7a depicts two microstructural regions: one for the MIPS-HAp coating part not exposed to the SBF solution, and the other for the part exposed to the SBF solution. As shown by the photomicrograph taken at a higher magnification and presented in Figure 5.7b, a significant change of the MIPS-HAp coating morphology happened after 1 day of immersion in SBF. The left side of the SEM photomicrograph (Figure 5.7b) showed the formation of fine-needle-like structures, as has been also observed for dense MAPS-HAp coatings [3] exposed to SBF solution, while the right side was almost unaffected. The microstructure revealed on the left side of Figure 5.7b represents the deposition of the carbonated apatite layer. Thus, these SEM photomicrographs corroborate the evidence of carbonated apatite formation, which was also confirmed from the FTIR spectra and XRD data, as mentioned earlier. It has been opined [9, 10] that the calcium and phosphate ions from the supersaturated SBF solution are consumed for spontaneous growth of crystalline apatite once it has nucleated and formed.

The EDX spectrum data of the MIPS-HAp coatings after the first day of immersion in the SBF solution are shown in Figure 5.8. The data showed peaks corresponding to the presence of the Ca and P phases. The Ca/P ratio was around 1.68. This Ca/P ratio was almost close to what would be a stoichiometric Ca/P ratio of 1.667. In the case of denser HAp coatings, other researchers [340, 341] have reported that a nonstoichiometric Ca/P ratio of 1.4 to 1.8 could also happen due to the formation of carbonated apatite and other Ca-P compounds. Interestingly, the minute presence of Mg, Na, and Cl [3, 4, 9, 13, 25] in the compositional characteristics reflects compositional characteristics that are the same as those reported [9, 26] for bone mineral. Further, the conventional SEM photomicrograph of the MIPS-HAp coatings after 4 days of immersion in the SBF solution is shown in Figure 5.9a, b. This photomicrograph shows a complete conversion of microstructure. Some distinct dissolution sites were also observed in the precipitated needle-like structure shown in the SEM photomicrograph obtained at a higher magnification (Figure 5.9b). The deposition of the apatite layer and its dissolution took place simultaneously, as expected [3, 16]. This aforesaid phenomenon

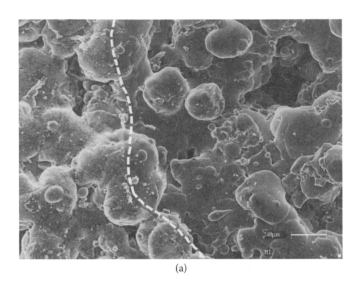

(a)

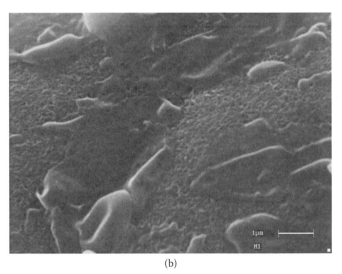

(b)

FIGURE 5.7
(a) Conventional SEM photomicrographs of the MIPS-HAp coatings after 1 day of immersion in the SBF solution showing a distinct region, e.g., modified region (left side), and an as-sprayed unchanged region. (b) Higher-magnification view of the modified microstructure showing the needle-like nanostructured apatite layer formation.

(i.e., simultaneous deposition and dissolution) also corroborated the data presented in Figure 5.3a, which showed the evidence of both deposition and dissolution.

After 14 days of immersion in the SBF solution, deposition was observed over the entire surface of the MIPS-HAp coating (Figure 5.10a), although

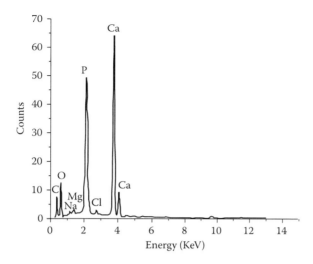

FIGURE 5.8
The EDX spectra of the MIPS-HAp coatings after 1 day of immersion in the SBF solution. (Reprinted from Dey and Mukhopadhyay, *International Journal of Applied Ceramic Technology*, 11: 65–82, 2013. With permission from American Ceramic Society and Wiley.)

some submicron dissolution sites had also appeared (Figure 5.10b). The microcracks, which are characteristically produced during the plasma spray deposition process, are now covered by the deposited fine-nano-needle-like network structures. It may be recalled that exactly similar features had also been observed from the photomicrographs presented earlier in this chapter and obtained from the corresponding E-SEM studies. It is interesting to mention that for denser MAPS-HAp coatings, similar observations were reported by Gross and Berndt [2].

The data on the Ca/P ratio as a function of the number of days passed after immersion in the SBF solution are summarized in Table 5.2. If we have a look at the data, they tell us that the Ca/P ratio was almost stoichiometric after the first day of SBF immersion. However, on the fourth and seventh days it was marginally lower, i.e., Ca/P ~ 1.59 and 1.57, respectively. It has already been mentioned that the formation of OCP or other Ca-P compounds might reduce the Ca/P ratio [4, 7]. Based on the present experimental data, it may also be suggested that small amounts of magnesium and sodium can substitute for calcium in the HAp lattice [25]. If and when this happens, that would lower the Ca/P ratio, as was also noted in the present experimental data (Table 5.2). In contrast, after the 14th day of immersion, the Ca/P ratio had increased to 1.82. This observation can be rationalized in terms of the fact that when carbonate substitutes for phosphate, HAp is transformed into carbonated apatite (CHAp). The enhancement in the Ca/P ratio matches with such a picture.

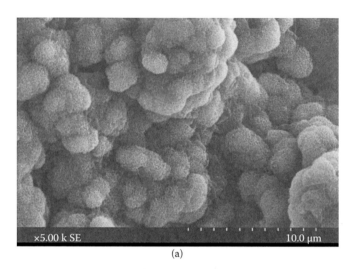

(a)

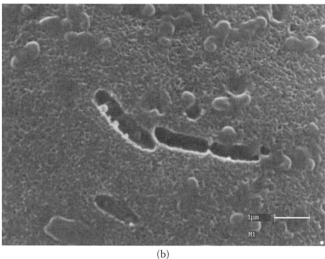

(b)

FIGURE 5.9
(a) Conventional SEM photomicrograph of the MIPS-HAp coatings after 4 days of immersion in the SBF solution showing the entire conversion of the microstructure. (b) The higher magnification shows the detail. (Reprinted from Dey and Mukhopadhyay, *International Journal of Applied Ceramic Technology*, 11: 65–82, 2013. With permission from American Ceramic Society and Wiley.)

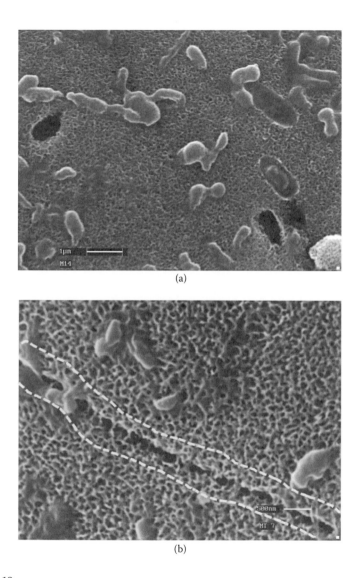

(a)

(b)

FIGURE 5.10
Conventional SEM photomicrographs of the MIPS-HAp coatings after 14 days of immersion in the SBF solution showing (a) both deposition and dissolution sites and (b) healing of microcracks due to deposition of the apatite layer. (Reprinted from Dey and Mukhopadhyay, *International Journal of Applied Ceramic Technology,* 11: 65–82, 2013. With permission from American Ceramic Society and Wiley.)

5.6 Summary

The typical results of a systematic investigation on the dissolution of MAPS- and MIPS-HAp coatings following immersion in the SBF solution are discussed in the present chapter. As the amount of studies reported on the later type of coatings is far from significant, major importance has been devoted to the in-depth investigation of the dissolution of MIPS-HAp coatings. The first emphasis has been placed on the different types of physical, chemical, and microstructural changes that can happen following immersion. Simultaneously, it has been attempted to illustrate the signatures corresponding to these changes and the relevant experimental techniques that are to be utilized to pick up these signatures. Finally, it has been attempted to explain how and why these changes happen. The data obtained from the ICP-AES results confirmed that after the first day of immersion in the SBF solution, the dissolution of Ca and P was dominant. The additional results found from XRD, FTIR, SEM, E-SEM, and EDX experiments, however, confirmed that after the fourth and up to 14 days of immersion, the depositions of fine-needle-like nanostructured apatites were the dominant process. When the MIPS-HAp coatings deposited on biomedical-grade SS316L substrates were immersed in the SBF solution, the results obtained from the ICP-AES experiments confirmed that the leaching of toxic chromium and iron metal ions did not happen for up to 2 weeks of immersion time. These data would strongly suggest the reliability of MIPS-HAp-coated implants on SS316L for biomedical prosthesis applications.

The macromechanical properties of HAp coatings, e.g., bonding strength, shear strength, fatigue behavior, etc., will be discussed in Chapter 6. The issues governing HAp coatings' bonding strength, like microstructure, interfacial stress, post-heat treatment, etc., will be also described. The general guidelines to improve the bonding strength and its measurement procedure will be discussed in Chapter 6.

References

1. Kirkpatrick C. J. and Mittermayer C. 1990. Theoretical and practical aspects of testing potential biomaterials in vitro. *Journal of Materials Science: Materials in Medicine*, 1: 9–13.
2. Gross K. A. and Berndt C. C. 1994. In vitro testing of plasma-sprayed hydroxyapatite coatings. *Journal of Materials Science: Materials in Medicine*, 5: 219–224.
3. Ha S. W., Reber R., Eckert K. L., Petitmermet M., Mayer J., Wintermantel E., Baerlocher C., and Gruner H. 1998. Chemical and morphological changes of vacuum-plasma-sprayed hydroxyapatite coatings during immersion in simulated physiological solutions. *Journal of the American Ceramic Society*, 81: 81–88.

4. Lickorish D., Ramshaw J. A. M., Werkmeister J. A., Glattauer V., and Howlett C. R. 2004. Collagen-hydroxyapatite composite prepared by biomimetic process. *Journal of Biomedical Materials Research Part A*, 68: 19–27.
5. Huan Z., Chang J., and Zhou J. 2010. Low-temperature fabrication of macroporous scaffolds through foaming and hydration of tricalcium silicate paste and their bioactivity. *Journal of Materials Science*, 45: 961–968.
6. Fernandez J., Gaona M., and Guilemany J. M. 2007. Effect of heat treatments on HVOF hydroxyapatite coatings. *Journal of Thermal Spray Technology*, 16: 220–228.
7. Wen H. B., Liu Q., de Wijn J. R., and de Groot K. 1998. Preparation of bioactive microporous titanium surface by a new two-step chemical treatment. *Journal of Materials Science: Materials in Medicine*, 9: 121–128.
8. Zhang Q., Chen J., Feng J., Cao Y., Deng C., and Zhang X. 2003. Dissolution and mineralization behaviors of HA coatings. *Biomaterials*, 24: 4741–4748.
9. Takadama H., Kim H. M., Kokubo T., and Nakamura T. 2001. Mechanism of biomineralization of apatite on a sodium silicate glass: TEM-EDX study in vitro. *Chemistry of Materials*, 13: 1108–1113.
10. Gamble J. 1967. *Chemical anatomy, physiology and pathology of extracellular fluid*. Cambridge, MA: Harvard University Press, 1.
11. Kim H. M., Himeno T., Kawashita M., Kokubo T., and Nakamura T. 2004. The mechanism of biomineralization of bone-like apatite on synthetic hydroxyapatite: an in vitro assessment. *Journal of the Royal Society Interface*, 1: 17–22.
12. Stanciu G. A., Sandulescu I., Savu, B., Stanciu S. G., Paraskevopoulos K. M., Chatzistavrou X., Kontonasaki E., and Koidis P. 2007. Investigation of the hydroxyapatite growth on bioactive glass surface. *Journal of Biomedical and Pharmaceutical Engineering*, 1: 34–39.
13. Bharati S., Sinha M. K., and Basu D. 2005. Hydroxyapatite coating by biomimetic method on titanium alloy using concentrated SBF. *Bulletin of Material Science*, 28: 617–621.
14. Yang L., Hedhammar M., Blom T., Leifer K., Johansson J., Habibovic P., and Blitterswijk C. A. V. 2010. Biomimetic calcium phosphate coatings on recombinant spider silk fibres. *Biomedical Materials*, 5: 045002.
15. Grigorescu S., Ristoscu C., Socol G., Axente E., Feugeas F., and Mihailescu I. N. 2005. Hydroxyapatite pulsed laser deposited thin films behaviour when submitted to biological simulated tests. *Romanian Reports in Physics*, 57: 1003–1010.
16. Khor K. A., Li H., Cheang P., and Boey S. Y. 2003. In vitro behavior of HVOF sprayed calcium phosphate splats and coatings. *Biomaterials*, 24: 723–735.
17. Zhao Q. Y., He D. Y., Li X. Y., and Jiang J. M. 2009. In vitro study of microplasma sprayed hydroxyapatite coatings in simulated body fluid. *Advanced Materials Research*, 79–82: 815–818.
18. Zhao Q., He, D., Zhao L., and Li X. 2011. In-vitro study of microplasma sprayed hydroxyapatite coatings in Hanks balanced salt solution. *Materials and Manufacturing Processes*, 26: 175–180.
19. Dey A. and Mukhopadhyay A. K. 2013. In-vitro dissolution, microstructural and mechanical characterizations of microplasma sprayed hydroxyapatite coating. *International Journal of Applied Ceramic Technology*, 11: 65–82.
20. Dey A. 2011. Physico-chemical and mechanical characterization of bioactive ceramic coating. PhD dissertation, Indian Institute and Engineering Technology (formerly Bengal Engineering and Science University), Shibpur, Howrah, India.

21. Kokubo T. and Takadama H. 2006. How useful is SBF in predicting in vivo bone bioactivity? *Biomaterials*, 27: 2907–2915.
22. ISO 10993-14. 2001. Biological evaluation of medical devices. Part 14. Identification and quantification of degradation products from ceramics. International Organization for Standardization (ISO).
23. Khor K. A., Li H., and P. Cheang. 2003. Characterization of the bone-like apatite precipitated on high velocity oxy-fuel (HVOF) sprayed calcium phosphate deposits. *Biomaterials*, 24: 769–775.
24. Chakraborty J., Sinha M. K., and Basu D. 2007. Biomolecular template induced biomimetic coating of hydroxyapatite on an SS316L substrate. *Journal of the American Ceramic Society*, 90: 1258–1261.
25. Habibovic P., Barrere F., vanBlitterswijk C. A., de Groot K., and Layrolle P. 2002. Biomimetic hydroxyapatite coating on metal implants. *Journal of the American Ceramic Society*, 85: 517–522.
26. LeGeros R. Z. and LeGeros J. P. 1993. *An introduction to bioceramics*. Singapore: World Scientific Publishing, 139.

6

Macromechanical Properties
of Hydroxyapatite Coating

6.1 Introduction

It is well known that bioactive coatings are implanted in vivo. They must have good adhesion with the substrate. Otherwise, they may fail during in-service condition. This is an undesired situation because it would obviously require revision surgery. It is from this perspective that the macromechanical properties of a given bioactive ceramic coating, e.g., HAp coating, need to be thoroughly evaluated and analyzed in relation to its microstructure. This is exactly what we plan to do in this chapter.

Now the mechanical properties that are of importance are quite a few. The first important property that affects the quality of adhesion is the bonding strength of the coating with the substrate. It also needs to be appreciated that there are situations when a coated implant undergoes relative motion under load. Under such situations the structural integrity is determined by the shear strength. So, the shear strength of the HAp coating shall be discussed. Further, the most common situation that is encountered by a coated implant is that it is sequentially loaded and unloaded, for instance, during walking, running, jogging, or any other body movement that involves alternate placement of body weight on different legs. This situation is similar to fatigue loading. Therefore, we shall make an attempt to discuss the mechanical fatigue behavior of HAp coatings. Finally, it is to be further appreciated that it will be utopian thinking to imagine a crack-free bioactive ceramic coating. If and when they are prospectively loaded, e.g., in tension, such microstructurally omnipresent cracks can certainly grow to such a critical dimension as to cause through-thickness propagation, leading to eventual failure of the coating itself. Since this will be the most undesirable situation to encounter in practice, we shall discuss the crack propagation behavior under the three-point bending loading condition.

To develop these ideas in a plausible manner, therefore, we need to look into the factors that govern the performance of a HAp coating, the interface related issues, what is really meant by the bonding strength of a coating and what are the means to evaluate the same, how one can improve the bonding strength of a coating, and what are the other important parameters that can affect the bonding strength.

Such parameters may include, but are not necessarily limited to, the influences of adhesives, coating microstructure, vacuum heat treatment, interfacial stresses, substrate holding arrangement, failure modes, relative humidity, and finally, dissolution behavior on the bonding strength. Such a discussion will prepare us to understand the perspective of measurement techniques other than the American Society for Testing and Materials (ASTM) method and the global development of HAp coatings by processes other than the conventional plasma spraying processes.

This backdrop will help us to understand the issues of utmost importance in evaluation and analysis of the bonding strength of MIPS-HAp coatings, as well as to develop a comparative idea of the bonding strength of MIPS- and MAPS-HAp coatings. Next, we shall discuss the role of residual stress in affecting the bonding strength, followed by a general discussion on shear strength and pushout strength of coatings. A detailed discussion will be made in relation to crack propagation behavior under three-point bending loading of a MIPS-HAp coating prior to taking up, finally, the picturization of the fatigue scenario in such coatings. Thus, the subsections below will provide a sequential discussion on the issues highlighted above.

6.2 What Governs HAp Coating's Performance?

The major characteristics that affect the ultimate performance of a HAp coating include [1–5] (1) coating thickness, (2) surface conditions (roughness and cleanliness) of the metal substrate, (3) mechanical strength of the coating, (4) porosity of the coating, (5) chemical purity of HAp after spraying, (6) crystallinity of the HAp material, and (7) dissolution properties of the coating. Lemons [2] opined that these aforesaid factors determine the biological longevity of the coating, and thereby influence the clinical outcome. A thin HAp coating of about 50–70 μm with full density was demonstrated to give high bonding strength with the substrate and to overcome resorption [3].

It has been further proposed that a HAp purity of >95% and HAp crystallinity of >70% after spraying were conducive to obtain good bonding strength [1]; however, the exact quantitative methods used to evaluate HAp content and coating crystallinity were not clearly mentioned [3]. Therefore, these results clearly demonstrate that the optimum values for chemical purity and crystallinity have not yet been unequivocally established.

6.3 Interface Issues

In a study by Spivak et al. [4], the failure of the bone-HAp coating during interfacial tensile strength test occurred consistently at the HAp coating–metal substrate interface. This indicated clearly that the challenge was, and in fact still is, to develop reliable bonding between the HAp coating and the metallic substrate. Therefore, emphasis must be given to the promotion of bonding at the interface between the HAp coating and the metallic substrate.

6.4 Bonding Strength and Methods of Measurements

The true bonding strength of the plasma sprayed coatings was proposed to be a manifestation of the mixture of the cohesive bonding strength between the lamellar layers themselves and the adhesive bonding strength between the coating and the substrate [5]. Conventionally, the bonding strength of the HAp coating to the metallic substrate was tested using the standard bonding test method as per ASTM standard C-633-01 [6].

For this test, the HAp coatings of thickness 100–300 µm were sprayed on cylindrical stubs of diameter 24.5 mm (Figure 6.1). The length of substrate should also be 24.5 mm, or approximately 1 in. Next, the coated stub was joined to another uncoated stub with a commercially available adhesive tape or adhesive glue (Figure 6.2). The stubs were then mounted on a tight-fixing fixture and kept in an oven for several hours. Thereafter, the test was carried out by universal testing machine under ambient conditions.

FIGURE 6.1 (See color insert.)
A typical image of the MIPS-HAp coating on SS316L cylindrical substrate (both diameter and length of ~25 mm) of different thicknesses, e.g., 100, 200, and 300 µm (on left side onward), and grit-blasted uncoated SS316L cylindrical substrate before the coating (on extreme right side image).

FIGURE 6.2 (See color insert.)
Arrangements before bonding strength measurement: two identical SS316L cylindrical stubs (e.g., top one coated and bottom one uncoated) joined with adhesive tape and cured in oven for several hours.

The bonding strength of the coating was obtained by dividing the critical load at failure by the coated area.

6.5 What Are General Guidelines to Improve Bonding Strength?

Literature reports [5–28] show that (1) a denser microstructure [5], (2) a thinner HAp coating (less than 100 µm) [5, 25, 26], and (3) a post-deposition heat treatment in vacuum all resulted in higher bonding strength [27].

6.6 Other Important Parameters

The other important parameters that affected the bonding strength of the HAp coatings as reported in literature are:

1. The surface conditions (roughness and cleaning) of the metal substrate
2. The properties (e.g., viscosity, adhesive strength, and shrinkage rate) of the glue employed
3. The extent of penetration of the glue into the imperfections (e.g., porosities and microcracks) of the coating

6.7 Influence of Adhesive

In a study by Filiaggi and co-researchers [28], the effect of the properties of adhesive glue on the bonding strength was evident. They used special glue that had high viscosity, took very little time to set, but had a weaker strength value, and they finally obtained a typically lower magnitude of the bond strength (~7 MPa).

On the other hand, in agreement with the work of Filiaggi et al. [28], a similar low value of bonding strength was reported (e.g., ~9 MPa) even when a stronger glue was employed [7]. Therefore, this level of strength data might represent mostly the interlamellar cohesive strength due to the fact that the glue was of a quick setting type. Munting et al. [25] achieved a bonding strength of 60–70 MPa for a thin coating (50 μm); however, the validity of the test was questionable and the ultra-high bonding strength reported could have reflected the strength of the glue used.

6.8 Influence of Microstructure

In an attempt to explain the mechanism that governs the bonding strength of HAp coatings, Wang et al. [29] showed that the HAp coatings with a denser microstructure had a higher extent of HAp powder melting, and hence a higher bonding strength. In this work, the HAp coatings with the densest structure exhibited a higher bonding strength (~30 MPa) than that (~23 MPa) of the most porous HAp coating. They also concluded that the HAp coating that exhibited a relatively higher extent of the HAp powder melting during the coating process also had the higher content of impurity phases, lower crystallinity, and higher Ca/P molar ratio. The reverse was true for the HAp coating that had a relatively lower extent of the HAp powder melting during the coating process.

Yang and Chang [30] have opined that an increase in the volume percent open porosity could strongly degrade the cohesive bonding of both A-HAp coatings prepared using the fixed holder and B-HAp coatings prepared using the rotational holder. The coatings produced using the fixed holder (i.e., the A-HAp coatings) exhibited a rough surface morphology, higher crystallinity and porosity content with nonuniformity of porosity distribution, as well as a higher residual stress. The nonuniformity in porosity distribution in A-HAp coatings was attributed to have played a major role in lowering the bonding strength through a strong degradation of the cohesive strength of the coatings.

6.9 Influence of Vacuum Heat Treatment

Recently, Yang [8] measured the bonding strength between the plasma sprayed-HAp coatings and substrate after heat treatment at various temperatures in vacuum (e.g., better than 10^{-5} torr). The bonding strength of the as-sprayed coating was about 29 MPa, which increased with increasing temperature up to 600°C to about 42 MPa, but above 600°C and up to 900°C, it degraded sharply to about 20 MPa, which was even lower than that of the as-sprayed HAp coating. It was suggested that the bonding strength of the HAp coating on Ti substrates was improved through the effects of the crystallization and the sintering-induced densification of the HAp coating when the post-spraying vacuum heat treatment temperature was below 600°C.

In another study, Yang and Chang [30] concluded that when the post-spraying vacuum heat treatment temperature was below 600°C, the compressive residual strains of HAp coatings were released. According to them [30], this also helped to improve the bonding strength of the HAp coatings. However, at temperatures above 600°C, the compressive residual stresses of the HAp coatings reduced the interfacial adhesive force of the HAp coating on the substrate, and as a result, the bonding strength degraded [30].

6.10 Role of Interfacial Stress

In a related study, it has been found [30] that the in-plane compressive residual stress would induce through-thickness tensile stress acting in the direction normal to the interface of the HAp coating and the Ti substrate. The role of this stress was to neutralize the bonding force between the coating and the substrate. It also reduced the interfacial adhesive force. Now these interfacial adhesive forces were actually the forces that degraded the bonding strength between the coating and the substrate. Therefore, the magnitude of the bonding strength increased with the increasing temperatures when the post-spraying vacuum heat treatment temperatures were raised to ≤600°C. These researchers opined that the reasons for this strength improvement were not only the sintering-induced densification of the HAp coating, but also the higher magnitude of the interfacial adhesive force, which resulted from the release of the compressive residual strain of the as-sprayed HAp coating through the post-spraying vacuum heat treatment.

It was interesting to note that when the post-spraying vacuum heat treatment temperature was raised above 600°C, the coefficients of thermal expansion (CTE) of the HAp coating were smaller than that of Ti alloy substrate. As a result of this process, a larger compressive residual strain occurred, and it increased with the increasing temperatures. Consequently, the bonding strength was weakened with heat treatment in vacuum at temperatures higher than 600°C.

6.11 Role of Substrate Holding Arrangements

The experimental results [7, 30] showed that the bonding strengths of the HAp coatings obtained using a fixed substrate holder were significantly lower than those of the HAp coatings obtained using a rotational substrate holder [29, 30]. These observations could be rationalized in terms of the variation in microstructure and the resultant porosity.

6.12 Failure Mode and Related Issues

Yang and Chang [30] showed that the area fraction of adhesive failure for each series of HAp coatings was apparently influenced by the compressive residual stress. In turn, the area fraction of adhesive failure of the specimens was apparently correlated with the bonding strength in the HAp coatings obtained using a fixed substrate holder and the HAp coatings obtained using a rotational substrate holder. For each series of HAp coatings, it was also shown that the larger the area that failed adhesively, the lower was the bonding strength of both HAp coatings.

In another study, Yang and Chang [7] tried to understand whether residual stress might be an important factor influencing the bonding strength between a ceramic coating and the metallic substrate. Six numbers of HAp coatings were produced by plasma spraying on Ti-6Al-4V substrates of different initial temperatures in the presence of various cooling media that had been utilized during the plasma spraying processes. The HAp coating with the lowest residual stress exhibited a higher bonding strength (e.g., ~9 MPa) than those of the others, while the HAp coatings obtained with high residual stress due to the deliberate absence of any cooling gas displayed the lowest bonding strength (e.g., ~2 MPa).

The observation that the specimen with the higher residual stress exhibited lower bonding strength at the interface of the HAp coating on Ti-6Al-4V substrate can be explained in terms of the compressive residual stress in the coating, which could have easily caused the coating to delaminate from the substrate.

Further, from the fractographs of adhesive tests of the HAp coatings, the failure modes revealed both cohesive and adhesive failures of the coating. After the adhesion test, the adhesive failure area of HAp coating that displayed the lowest bonding strength was about 68%, while the HAp coating that displayed the highest bonding strength revealed only about 23% of the adhesive failure area.

In general, HAp coatings having the dense microstructures obtained through plasma spraying with high power possessed a higher bonding strength than coatings having such microstructures obtained through plasma spraying with lower power. This was not only due to the differences in adhesive strengths of the related HAp coatings.

The magnitude for bonding strength actually reflects the combination of the adhesive strength (i.e., the strength/weakness of the interface between the coating and the substrate) and the cohesive strength (i.e., the strength/weakness within the coating layers themselves) of a coating. Of these two aspects, the cohesive strength is dominated by the coating structure, i.e., crystallinity, porosity, crack and lamellar texture, etc. In contrast, the adhesive strength might be affected by, e.g., the coating microstructure, the residual stress, and the surface roughness of the substrate.

6.13 Influence of Humidity

Ding et al. [18] studied the variation of the bond strength of monolithic HAp coating with time of storage in a humid environment. In conditions of low (e.g., <30%) humidity, the bond strength did not change with time. However, when exposed to high (e.g., >95%) humidity, the monolithic HAp coating on Ti-6Al-4V substrates significantly lost its bonding strength.

Further, with increasing storing time in high-humidity conditions, e.g., after 30 days, a continuous degradation from about 60 MPa in the as-sputtered samples to about 50 MPa was noted. Thus, a reduction of about 17% of the original bonding strength happened in highly humid conditions.

It was suggested that the bonding strength degradation of HAp coating came mostly from the water attack. When water from the atmosphere was absorbed throughout the coating by capillary action, it infiltrated into the inner portion of the HAp coating through the narrow structural imperfections and came into contact with the deeper portions of the coating [31].

The interaction of absorbed water and coating finally resulted in the weakening of the bonding strength between the HAp coating and the substrate [32]. These results imply the need for tremendous care that should be taken in storing HAp-coated Ti-6Al-4V implants.

6.14 Influence of the Dissolution Behavior

Yang and co-researchers [20, 33] showed that the HAp coatings having the most dense microstructure before immersion in simulated body fluid (SBF) had the highest bonding strength and the highest interlamellar cohesive strength. HAp coatings with denser microstructure would lead to lesser instability. This rationale could explain why the most dense HAp coating dissolved only slightly, and consequently, only a 30% reduction of original bond strength (e.g., from ~31 MPa to ~21 MPa) was measured even after 4 weeks of immersion [33].

On the other hand, in another related study [20], a higher bonding strength degradation (e.g., from ~30 MPa to ~17 MPa after 1 week of immersion in SBF) resulted from serious dissolution of the HAp coating, which had a highly porous microstructure. Nevertheless, this trend was followed by an abnormal increase in strength data (~24 MPa) measured during the last 3 weeks of immersion in SBF. This abnormal rise in bonding strength was suggested to be a contribution of the glue that penetrated the coating microstructure. A typical range of about 25–33%, or even higher, reduction in bond strength was measured.

These researchers [20, 33] opined that in evaluating the dissolution behavior of the HAp coatings, the microstructure, rather than coating crystallinity, seemed to be the most important variable. In addition, as the amorphous component of HAp coatings could crystallize in SBF easily, the initial crystallinity of the coating might easily be rendered to be a less important factor of influence. Evidence of the dissolved morphologies of HAp coatings that weakened the interlamellar strength had accounted for the degraded bond strength.

6.15 Bonding Strength Measurements by Techniques Other than ASTM C-633

The details of the bonding strength values reported by several researchers in literature for HAp and HAp composite coatings on different substrates deposited by several coating techniques are presented in Table 6.1 [5, 7–24, 30, 34–37].

A modified ASTM C-633 method was employed to measure the bonding strength of the HAp or HAp composite coating [16, 17]. However, the maximum bonding strength value (~100 MPa) of HAp coating by the sol-gel technique was reported in one case [14]. In this case, bonding strength was evaluated by an ultra-microhardness indentation method. Other methods, e.g., pull-off test [18], scratch test [34], etc., were also employed to measure the bonding strength of the HAp and HAp composite coatings.

TABLE 6.1

Literature Status on Bonding Strength of HAp or HAp Composite Coatings

Coating/Substrate	Depostion Process	Bonding Strength[a] (MPa)	Reference
HAp/Ti-6Al-4V	MAPS	~8–14 (4–12 weeks in SBF)	[5]
HAp/Ti-6Al-4V	MAPS	2–9	[7]
HAp/Ti-6Al-4V	MAPS	20–42	[8]
HAp-BG/Ti-6Al-4V	MAPS	61–64	[9]
HAp/Ti	PS	13–17	[10]
HAp composite/Ti-6Al-4V	MAPS	45–50	[11]
YSZ-HAp/SS	SG	17–32 (S.S.)	[12]
HAp/Ti-6Al-4V	SG	>14	[13]
HAp/Ti-6Al-4V	Aero-SG	100 (UMI)	[14]
HAp-Na-Ti/Ti-6Al-4V	PS	27–38	[15]
HAp/BG, GC, H_d, Ti	B	2–30 (modified ASTM-633)	[16]
HAp/Ti-6Al-4V	PLD	>58 (modified ASTM-633)	[17]
HAp & HAp-Ti (FGC)/ Ti-6Al-4V	MS	(Pull-off test, Sebastian system) ~60 ~50–55 (10–30 days in SBF) ~60–82 (FGC, increase Ti) ~55–83 (FGC 10–30 days in SBF)	[18]
HAp/Ti-6Al-4V		~ 35 <10 (Bone to HAp)	[19]
HAp/Ti-6Al-4V	MAPS	~ 23–30 ~17–30 (in SBF 1–4 weeks) ~15–17 (in SBF 12–24 weeks)	[20]
HAp/Ti-6Al-4V	MAPS	~27–31 ~18–30 (in SBF 1–4 weeks)	[21]
HAp, HAp-glass/zirconia	Dip coating	~22 ~35–40 (HAp-G)	[22]
HAp/C-C	MAPS	4–9 (S.S.)	[23]
HAp		~1.6 (bone to HAp) (after 6 weeks in rabbit)	[24]
HAp/Ti-6Al-4V	MAPS	9–26	[30]
β-TCP, FHAp/Ti-6Al-4V	SG	~500 mN (scratch test, max load at which coating peels off)	[34]
HAp, HAp-Ti-6Al-4V/ Ti-6Al-4V	PS	~27 (HAp-TA) ~18 ~27–16 (8 weeks in SBF)	[35]
HAp-YSZ/Ti-6Al-4V	MAPS	23–30	[36]
HAp/SS316L	MIPS	13	[37]

TABLE 6.1 (continued)

Literature Status on Bonding Strength of HAp or HAp Composite Coating

Note: MAPS = macroplasma spraying, MIPS = microplasma spraying technique, SG = sol-gel technique, PLD = pulsed laser deposition, B = biomimetic technique, p = percentage of porosity, UMI = ultra-microindentation, SBF = simulated body fluid, C-C = carbon-carbon composite, YSZ = yttria-stabilized zirconia, FGC = functionally graded coating, S.S. = shear strength of the coating, SS = stainless steel, FHA = fluoridated hydroxyapatite, BG = bioglass, GC = glass ceramic, and H_d = dense hydroxyaptite.

[a] Bonding strength of the coating tested by ASTM C–633.

6.16 HAp Coatings Developed by Other Coating Processes

Other than the conventional plasma spraying, either macroplasma spraying (MAPS) or microplasma spraying (MIPS) based, HAp or HAp composite coatings have been prepared by a variety of methods, e.g., sol-gel [12–14, 34], pulsed laser deposition [17], biomimetic [16], dip coating [22], etc., on different substrates, for instance, stainless steel [12], titanium alloys [13, 14, 16, 34], titanium [16], ceramics, e.g., bioglass [16], zirconia [22], or glass ceramic [84].

In the case of coatings prepared by the sol-gel technique, a shear strength of 17–32 MPa [12] and bonding strength of at least 14 MPa [13] have been typically recorded. Even nanoscratch tests have been employed to opine that at least up to a load of 500 mN, the coating does not peel off [34].

In the case of coatings prepared by the pulsed laser deposition technique, it is reported that a so-called modified ASTM C-633 technique gave a bonding strength of at least 58 MPa [17]. Biomimetic coatings, however, record a much wider range of bonding strength, e.g., 2–30 MPa, when tested by the modified ASTM C-633 technique [16]. Similarly, in the case of the coatings prepared by the dip coating technique, the bonding strength measured by the conventional method (ASTM C-633) gave a value of 22 MPa for pure HAp and a slightly higher value (e.g., 35–40 MPa) for the HAp composite [22].

6.17 Bonding Strength of MIPS-HAp Coatings

The present authors and their coworkers [37] developed HAp coatings by the MIPS process, which has been discussed in detail in the previous chapters. Now we shall concentrate on its bonding strength properties. It will be pertinent to remember that the MIPS-HAp coatings are highly porous in nature. This is, in fact, beneficial for osseointegration, as well as bioresorption.

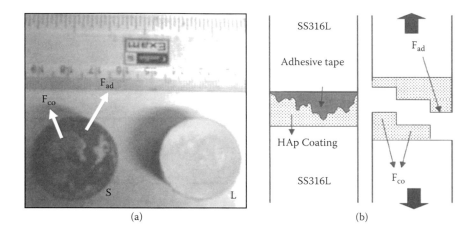

FIGURE 6.3
(a) Fractographs after bonding strength test on the MIPS-HAp coating showing the adhesive fracture (F_{ad}) and cohesive fracture (F_{co}) (L = loading stub, S = substrate stub). (b) Schematic representation of F_{ad} and F_{co}. (Reprinted from Dey et al., *Journal of Thermal Spray Technology*, 18: 578–592, 2009. With permission from Springer.)

However, too much porosity in the microstructure may sometimes cause total delamination of the coatings. Therefore, it is really crucial to study the bonding strength of such porous MIPS coatings.

Dey and his coworkers [37] recently reported average bonding strength of MIPS-HAp coatings as ~12.6 ± 0.31 MPa. After the bonding strength test in the universal testing machine in tensile mode, a typical loading stub, L, and the substrate stub, S, are shown in Figure 6.3a. The failure modes of the coating involved both cohesive failure zones (F_{co}), where fracture occurred mainly in between the ceramic splats themselves (Figure 6.3a), and adhesive failure zones (F_{ad}), where fracture occurred (Figure 6.3b) in between the ceramic splats and substrate. It is obvious that the more cohesive failure that occurs, the higher is the bonding strength of the coating. If the scenario is opposite, then splat-to-splat adhesion is poor. This may lead to production of debris from the coating itself. The cohesive failure takes place when the two successive layers of splats tear and make a route of intersplat fracture during the bonding strength measurement. Bonding strength is straightaway reliant on the mode of fracture. Therefore, it is expected that when the proportion of adhesive failure is more, there would be better adhesion between the coating and the substrate [7].

For the MIPS-HAp coatings, the present authors measured the bonding strength as ~13 MPa, which is adequate if we look into the reported literature values, which spread in a wide range, e.g., ~2–30 MPa. We have to keep in mind that the literature data are only available for the macroplasma-deposited HAp coating. Now, for the point of discussion, is this really encouraging data? Or is it really poor data? If we recall our earlier discussion regarding the bonding strength of the plasma sprayed coating in general, it mainly depends on a lot of factors, for example:

1. Porosity of the coating
2. Thickness of the coating
3. Post-spraying heat treatment of the coating
4. Choice of the substrate
5. Residual stress

It has been also reported that superior values of bonding strength were typically associated with MAPS-HAp coatings with (1) highly dense microstructure (385 pl chk n chng), (2) less coating thickness [25, 26, 38, 39], and (3) post-heat teat treatment in vacuum [8, 27]. In contrast, a comparatively lower bonding strength of MAPS-HAp coating on Ti alloy (e.g., ~7 and ~2–9 MPa) was also reported by Filiaggi and coworkers [28] and Yang and Chang [7]. Hence, it is yet to be unambiguously established the control of which factors would totally ensure attainment of higher bonding strength of PS-HAp coatings in general, and MIPS-HAp coatings in particular. The present authors [37] opined that the moderate value of the bonding strength, i.e., 13 MPa in the MIPS-HAp coating, was presumably linked to the existence of porosity (e.g., ~20%) that was much higher than that (e.g., 2–5%) in a typical dense MAPS-HAp coating.

Further, if we consider porosity dependency of strength, it is expressed according to the empirical Spriggs' equation:

$$\sigma = \sigma_0 \exp(-bP) \qquad (6.1)$$

where σ and σ_0 are strength of a ceramic materials at porosity P and zero, P is the volume percent open porosity, and b is the preexponential factor. Thus, σ_0 represents the strength of the dense ceramic (e.g., $P = 0$). The higher end average of the reported data on bonding strength of a dense MAPS-HAp coating is about 30 MPa [8]. So, if we presume σ_0 was about 30 MPa, the strength of the porous coating would be predicted to be about ~16 MPa, following Equation (6.1), for an experimentally measured porosity of 20% [37]. Therefore, the predicted value of bonding strength is well matched with the experimentally measured bonding strength data reported by Dey et al. [37].

6.18 MAPS-HAp vs. MIPS-HAp Coatings

In MIPS-HAp coatings, a higher porosity level, e.g., 20% or more, is achieved than what could be attained by a HAp coating prepared by the macroplasma spraying technique without compromising the high degree of crystallinity. The high porosity level is actually requisite for several practical reasons. The pores basically allow tissue ingrowth, and thus anchor the prosthesis to the surrounding bone, thereby preventing the loosening of implants. Further, the distributed porosity ensures the blood and nutrition supply for the bone. Additionally, the porosity aids to locally relax the strain, and thus diminish the residual stress. This particular aspect has been discussed elaborately in Chapter 9. A large surface area-to-volume ratio in porous HAp coating could enhance the rate of bioresorption, which is an important requisite for bio-application of HAp coating [37, 40]. An important matter is that HAp coatings are required to be phase pure and highly crystalline in order to improve long-term stability in in vivo applications.

Although a high porosity might seem favorable for better osseointegration, a coating with much higher porosity is likely to possess extremely poor adhesion to the substrate [41, 42]. Consequently, the poor adhesion might delaminate the coating. In such a situation, part of the failed coating can give rise to (1) exposed metal surface and (2) debris comprising fractured coating. Under such a situation, metallic ions can leach out of the metallic stems. This is the most dangerous, undesirable condition. Besides, an excessive penetration of body fluids through a porous HAp coating might cause more dissolution of the coating material in the interface region, which would worsen the adhesion strength further. Hence, it is usually recommended that an appropriate trade-off between porosity and adhesion strength be devised [40, 41].

6.19 Effect of Residual Stress

Another important factor that affects the bond strength is residual stress. Residual stress in plasma sprayed coating arises from two main sources. First, there are the intrinsic or deposition stresses, which are generated during the cooling of molten, sprayed particles to the temperature of the substrate as the solidification process continues to its final stage. Second, differential thermal contraction stresses may arise during the post-fabrication cooling-down period.

The behavior and mechanism of residual stress generation is so complicated that it is always open to extensive independent study by itself [7]. Investigations of plasma sprayed ceramic coatings on metal substrates have suggested that maximum residual stress generally occurs at the interface of the coating and the substrate. Such a residual stress may initiate a debonding stress at the interface that often leads to adhesive failure of the HAp coating on the metallic substrate. The CTE of sintered HAp is different than that of the metallic substrates. Therefore, a high compressive or tensile stress state may prevail in the present HAp coating after cooling down from the elevated temperature attained during the plasma spraying process. Such a high value of residual stress may cause the HAp coating to buckle easily. If a situation as suggested above were prevailing, that would also induce a comparatively lower magnitude of bonding strength [37].

6.20 Shear Strength and Pushout Strength

The shear strength (S) at the plasma sprayed-HAp coating–bone interface was calculated by dividing the maximum pushout force by the total bone area in contact with the implant, by using the relationship $F/\pi DH$, where D is the diameter of the implant (4.76 mm), and H is the average cortex thickness.

Yang et al. [20] measured the interfacial shear strength between the MAPS-HAp coating on Ti-6Al-4V after in vivo implantation and the bone interface. The well-prepared fresh samples were placed in a testing jig, and using a universal testing machine, the implants were pushed out at a very slow loading rate of ~200 μm.min^{-1} from the surrounding bone. The force needed to loosen the implant was determined from the load vs. displacement curve. As such, for different periods of implantation, the average strength data of the coating having a relatively lower porosity (i.e., about 3.5%) were significantly higher than those of the coatings having a relatively higher porosity (i.e., about 9%). After 24 weeks of implantation, shear strength as high as about 17 MPa was observed for the denser coating. The failure site of implants after the pushout tests was conclusively identified to be at or near the HAp coating–bone interface and not at the Ti-6Al-4V–HAp coating interface.

However, several researchers have stated that a wide scatter existed in measured pushout strengths for the plasma sprayed HAp-coated titanium implant system [39, 43]. Due to huge variability in test conditions, a comparison between different data sets is yet to be made. This issue again highlights the urgent need for a standard of pushout test that could be followed internationally for the HAp-coated titanium-based alloys/SS316L implant systems. More recently, Dhert et al. [44] further pointed out several parametric factors that might influence the results of shear strength measurements:

1. Clearance of the hole in the support jig
2. Young's modulus of the implant
3. Cortical bone sample's own thickness
4. Final diameter of the implant

In the other related study, shear strength of the post-heat treated (vacuum, 2 h, 700°C) MAPS-HAp coating on C/C composite was measured to be ~7 MPa [23]. Recently, Balamurugan et al. [12] measured the shear strength of the composite coating of YSZ-HAp on SS316L substrate. The HAp coating was prepared by a sol-gel method. Here, the HAp coatings showed the lowest strength (~17 MPa), and the 50 vol% zirconia-reinforced HAp coating showed the highest strength (~32 MPa), illustrating that the strength of the HAp coatings can be significantly improved by the addition of zirconia. This information also highlights the need to look into various possible means of reinforcing the MAPS-/MIPS-HAp coatings with a view to achieve higher strength as well as other macro- and nanomechanical properties of relevance for prosthetic applications.

6.21 Three-Point Bending Test

To the best of the present authors' knowledge, so far the only data related to the three-point bending test on HAp-coated samples are reported from their own research group [37]. In that study, the deformation and crack propagation behavior of the MIPS-HAp coating on SS316L substrate was investigated by a three-point bending test with a conventional universal testing machine under a displacement-controlled mode at ambient test conditions. The HAp-coated surfaces on the SS316L bars are kept in the tensile side (Figure 6.4) [37]. The combined stress developed at the HAp-coated SS316L sample was calculated according to the following equation:

$$\sigma = 3\ PL/2\ wt^2 \tag{6.2}$$

where P is the applied load, L is the span length, w is the width, and t is the total thickness of the HAp-coated sample. Two types of samples had been

FIGURE 6.4 (See color insert.)
Arrangement for three-point bending test of MIPS-HAp coating on SS316L substrate (HAp coated surface is kept in the bottom, i.e., tensile side).

taken: (1) notched at the center of the SS316L bar and (2) unnotched. Further, some prefixed deflection had been given. Additional details may be found elsewhere [37]. In this work, the deformation or crack front behavior was studied by a high-speed CCD camera attached to an optical microscope.

The stitched micrographs of the crack path for both unnotched and notched HAp-coated SS316L samples are shown in Figures 6.5a, b and 6.6a–c, respectively [37]. The crack deflection and zigzag path of the crack across the MIPS-HAp coating were observed in the unnotched sample at prefixed cross-head displacements of 1.5 and 1.8 mm, respectively. However, no crack commencement was observed below the cross-head displacement of 1.5 mm. Therefore, at a lesser stress level, the coating was intact, i.e., undamaged. The cracks initiated at still higher loads in the HAp coatings as the tensile stress was further enhanced by means of increasing cross-head displacement.

The various modes of failure propagation, e.g., crack deflection, crack branching, and also local crack bridging over the coating surface, were

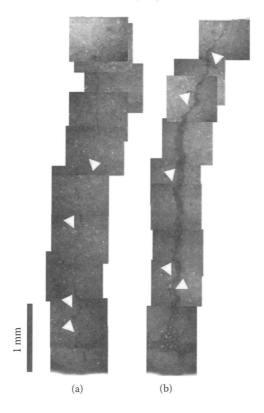

(a)   (b)

FIGURE 6.5
Stitched micrographs of crack propagation for the unnotched samples at prefixed cross-head displacements of (a) 1.5 and (b) 1.8 mm (crack deflection marked with arrowhead). (Reprinted from Dey et al., *Journal of Thermal Spray Technology*, 18: 578–592, 2009. With permission from Springer.)

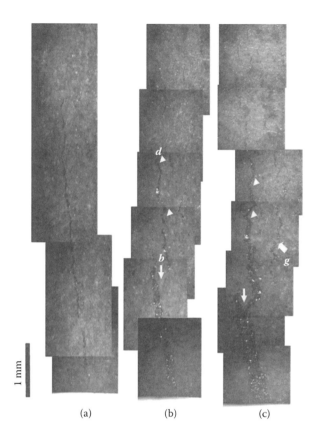

FIGURE 6.6
Stitched micrographs of crack propagation for the notched samples at prefixed cross-head displacements of (a) 1.5, (b) 1.75, and (c) 2 mm (d = crack deflection marked with arrowhead, b = crack branching marked as white arrow, g = local crack bridging marked as thick white arrow). (Reprinted from Dey et al., *Journal of Thermal Spray Technology*, 18: 578–592, 2009. With permission from Springer.)

observed in a notched sample at prefixed cross-head displacements of 1.5, 1.75, and 1.8 mm, respectively. The crack width at the beginning of the notched sample was much wider than those of cross-head displacements of 1.75 and 1.5 mm unnotched samples, as expected as the stress level increased. These observations may be rationalized in terms of differences between the direction of crack propagation and the local orientation of the splats in the coating microstructure. It is a fact that not much three-point bending test data obtained either by the interrupted test or by uninterrupted tests on either notched or unnotched samples are available on HAp coatings deposited by other techniques on either SS316L or Ti-6Al-4V or Ti. This information simply points out the need for much more database generation through precisely controlled and preplanned three-point bending test experimentations. There is no doubt that such an activity is very much deserved for the future development of the HAp coatings for biomedical implant applications in a much more reliable manner.

6.22 Fatigue Behavior

Fatigue properties also play an important role when the HAp coatings are used in an in vivo environment. More often, patients operated on with HApcoated implants experience different stress levels from their different physical movements. Therefore, mechanical integrity of HAp coatings should be adequate, as they should not fail in the body environment. The literature status on fatigue properties of HAp coating is summarized in Table 6.2 [45–49]. Lynn and Du Quesnay [46, 47] conducted fatigue tests at a frequency of 50 Hz and stress amplitude of 620 MPa on MAPS-HAp on Ti-6Al-4V. The thickness of the coating varied in the range of 25–150 μm; further, the coatings were heat treated at 400°C. As-sprayed coatings failed above 10^5 cycles, whereas the heat-treated coatings failed much earlier, e.g., only above 10^4 cycles. Further, the thinner (e.g., ~25–50 μm) coatings showed better performance than the thicker (~100–150 μm) HAp coatings. Lynn and Du Quesnay [46, 47] suggested that the fatigue property was highly dependent on the residual stress of the HAp coating. Fatigue tests have also been conducted at room temperature with the samples immersed in Ringer's solution [48, 49]. The results reported by Mukerjee et al. [49] showed that yield strength, yield strain, and elastic moduli depend on crystallinity of the coating after the fatigue test; however, the variations were random in nature. Further, they concluded that loss of HAp coating was a cumulative effect of fatigue as well as bacterial exposure. Lower crystalline HAp coatings showed a higher degree of coating loss.

Gledhill et al. [48] developed HAp coatings on Ti-6Al-4V by both the vacuum plasma spraying (VPS) and the DGS technique. Fatigue tests on these samples were conducted in Ringer's solution and at a frequency of 4 Hz over 1–10 million cycles. VPS-HAp coatings failed after 1 million cycles, whereas DGS-HAp coatings showed much superior performance and failed at a much later stage, e.g., after 10 million cycles. Zhang et al. [45] had conducted both static and dynamic fatigue tests. They found that thinner (~90 μm) coatings showed better performance than the thicker (~200 μm) one. However, fatigue tests on MIPS-HAp coatings are yet to be reported, suggesting that a priority research need exists in this domain.

TABLE 6.2

Literature Status on Fatigue Properties of HAp or HAp Composite Coating

Coating/Substrate: Processing Route	Details of Fatigue Test	Results	Reference
HAp/Ti-6Al-4V: MAPS	Static fatigue test (displacement rate = 0.3 mm.min^{-1}) Dynamic fatigue test (time: 10^7 cycle, stress ratio ~ 0.1%, frequency = 30 Hz) Shear strength test	Thinner (~90 μm) coating showed better performance than the thicker (~200 μm) one Shear strength = 25–40 MPa	[45]
HAp/Ti-6Al-4V: MAPS	Fatigue test (frequency = 50 Hz, stress amplitude = 620 MPa)	Thinner (~25–100 μm) coatings showed better performance than the thicker (~150 μm) one	[46]
HAp/Ti-6Al-4V: MAPS	Fatigue test (frequency = 50 Hz, stress amplitude = 620 MPa)	As-sprayed coatings fail above 10^5 cycles Heat treated at 400°C coatings fail above 10^4 cycles Thinner (~25–50 μm) coatings showed better performance than the thicker (~100–150 μm) coatings	[47]
HAp/Ti-6Al-4V: VPS, DGS	Fatigue test (in RT with Ringer's solution) (maximum tensile strain ~ 1%, frequency = 4 Hz, cycle: 1–10 million)	VPS-HAp coating fails: after 1 million cycles DGS-HAp coating fails: after 10 million cycles	[48]
HAp/Ti-6Al-4V: MAPS Crystallinity: 60.5% (A) 52.8% (B) 47.8% (C)	Fatigue test (in RT with Ringer's solution) (maximum tensile strain ~ 1%, frequency = 10 Hz, cycle: 5 million)	Yield strength: 923 MPa (A) 908 MPa (B) 950 MPa (C) Yield strain: 17.2% (A) 17.4% (B) 16.9% (C) Young's modulus: 8.8 GPa (A) 9.7 (B) 11.6 (C)	[49]

Note: MAPS = macroplasma spraying technique, VPS = vacuum plasma spraying technique, DGS = detonation gun spraying technique, COF = coefficient of friction, RT = room temperature.

6.23 Summary

We have discussed why macromechanical properties like bonding strength, shear strength, and fatigue properties of HAp coatings are particularly important when HAp coatings will be used in in vivo implantation. For HAp coatings processed by different methods, the comparison of bonding strengths and other properties has been provided. Of course, comparatively more importance has been devoted to plasma sprayed coatings in general, and the MIPS-HAp coatings in particular. The different factors influencing bonding strength, shear strength, and fatigue behavior have been discussed in detail. A thorough study of the relevant literature revealed that there is no fatigue study yet reported for MIPS-HAp coating. Further, during a three-point bending test, how HAp-coated metallic substrate will behave under the controlled prefixed stress level has been discussed. The behavior of crack front propagation during the three-point bending test has also been shown in this chapter.

In Chapter 7, the local mechanical properties like nanohardness, Young's modulus, and fracture toughness of HAp coatings will be discussed because these properties govern the contact-induced deformation resistance and stress transfer issues in actual load-bearing applications. The evaluations of nanohardness and Young's modulus by the advanced measurement technique, i.e., nanoindentation applied to HAp coatings, shall be elaborated. Further, the issue of statistical reliability of the experimental data on nanohardness and Young's modulus measured by the nanoindentation technique applied to the highly porous and heterogeneous MIPS-HAp coatings shall be critically examined in Chapter 7.

References

1. Geesink R. G. T. 1990. Hydroxyapatite-coated total hip prostheses. Two-year clinical and roentgenographic results of 100 cases. *Clinical Orthopaedics and Related Research*, 261: 39–58.
2. Lemons J. E. 1988. Hydroxyapatite coatings. *Clinical Orthopaedics and Related Research*, 235: 220–223.
3. Geesink R. G. T., de Groot K., and Christel P. A. K. T. 1987. Chemical implant fixation using hydroxylapatite coatings. *Clinical Orthopaedics and Related Research*, 225: 147–170.
4. Spivak J. M., Ricci J. L., Blumenthal N. C., and Alexander H. 1990. A new canine model to evaluate the biological response of intramedullary bone to implant materials and surfaces. *Journal of Biomedical Materials Research*, 24: 1121–1149.

5. Wang B. C., Chang E., Yang C. Y., Tu D., and Tasi C. H. 1993. Characteristics and osteoconduction of three different plasma-sprayed hydroxyapatite-coated titanium implants. *Surface and Coatings Technology*, 58: 107–117.

6. ASTM C-633-01. 2001. Standard test method for adhesion or cohesion strength of thermal spray coatings. In *Annual book of ASTM standards*.

7. Yang Y. C. and Chang E. 2001. Influence of residual stress on bonding strength and fracture of plasma-sprayed hydroxyapatite coatings on Ti-6Al-4V substrate. *Biomaterials*, 22: 1827–1836.

8. Yang Y. C. 2007. Influence of residual stress on bonding strength of the plasma-sprayed hydroxyapatite coating after the vacuum heat treatment. *Surface and Coatings Technology*, 201: 7187–7193.

9. Chen C. C., Huang T. H., Kao C. T., and Ding S. J. 2004. Electrochemical study of the in vitro degradation of plasma-sprayed hydroxyapatite/bioactive glass composite coatings after heat treatment. *Electrochimica Acta*, 50: 1023–1029.

10. Zheng X., Huang M., and Ding C. 2000. Bond strength of plasma-sprayed hydroxyapatite/Ti composite coatings. *Biomaterials*, 21: 841–849.

11. Ding S. J., Su Y. M., Ju C. P., and Lin J. H. C. 2001. Structure and immersion behavior of plasma-sprayed apatite-matrix coatings. *Biomaterials*, 22: 833–845.

12. Balamurugan A., Balossier G., Kannan S., Michel J., Faure J., and Rajeswari S. 2007. Electrochemical and structural characterisation of zirconia reinforced hydroxyapatite bioceramic sol-gel coatings on surgical grade 316L SS for biomedical applications. *Ceramics International*, 33: 605–614.

13. Weng W. and Baptista J. L. 1999. Preparation and characterization of hydroxyapatite coatings on Ti6Al4V alloy by a sol-gel method. *Journal of the American Ceramic Society*, 82: 27–32.

14. Manso-Silvan M., Langlet M., Jimenez C., Fernandez M., and Martinez-Duart J. M. 2003. Calcium phosphate coatings prepared by aerosol-gel. *Journal of the European Ceramic Society*, 23: 243–246.

15. Yang C. Y., Chen C. R., Chang E., and Lee T. M. 2007. Characteristics of hydroxyapatite coated titanium porous coatings on Ti-6Al-4V substrates by plasma sprayed method. *Journal of Biomedical Materials Research Part B*, 82: 450–459.

16. Kim H. M., Miyaji F., Kokubo T., and Nakamura T. 1997. Bonding strength of bonelike apatite layer to Ti metal substrate. *Journal of Biomedical Materials Research Part B*, 38: 121–127.

17. Garcia-Sanz F. J., Mayor M. B., Arias J. L., Pou J., Leon B., and Perez-Amor M. 1997. Hydroxyapatite coatings: a comparative study between plasma-spray and pulsed laser deposition techniques. *Journal of Materials Science: Materials in Medicine*, 8: 861–865.

18. Ding S. J., Lee T. L., and Chu Y. H. 2003. Environmental effect on bond strength of magnetron-sputtered hydroxyapatite/titanium coatings. *Journal of Materials Science Letters*, 22: 479–482.

19. Willmann G. 1999. Coating of implants with hydroxyapatite material connections between bone and metal. *Advanced Engineering Materials*, 1: 95–105.

20. Yang C. Y., Lin R. M., Wang B. C., Lee T. M., Chang E., Hang Y. S., and Chen P. Q. 1997. In vitro and in vivo mechanical evaluations of plasma-sprayed hydroxyapatite coatings on titanium implants: the effect of coating characteristics. *Journal of Biomedical Materials Research*, 37: 335–345.

21. Yang C. Y., Wang C., Chang E., and Wu J. D. 1995. Bond degradation at the plasma-sprayed HA coating/Ti-6Al-4V/alloy interface: an in vitro study. *Journal of Materials Science: Materials in Medicine*, 6: 258–265.
22. Kim H. W., Georgiou G., Knowles J. C., Koh Y. H., and Kim H. E. 2004. Calcium phosphates and glass composite coatings on zirconia for enhanced biocompatibility. *Biomaterials*, 25: 4203–4213.
23. Sui J., Li L. M. S., Lu Y. P., Yin L. W., and Song Y. J. 2004. Plasma-sprayed hydroxyapatite coatings on carbon-carbon composites. *Surface and Coatings Technology*, 176: 188–192.
24. Hong L., Hengchang X., and de Groot K. Tensile strength of the interface between hydroxyapatite and bone. *Journal of Biomedical Materials Research*, 26: 7–18.
25. Munting E., Verhelpen M., Li F., and Vincent A. 1990. Contribution of hydroxylapatite coatings to implant fixation. In *CRC handbook of bioactive ceramics*, Yamamuro T. L., Hench L., and Wilson J. (Eds.). Boca Raton, FL: CRC Press, 143–148.
26. de Groot K., Klein C. P. A. T., Wolke J. G. C., and de Blieck-Hogervorst J. M. A. 1990. Chemistry of calcium phosphate bioceramics. In *CRC handbook of bioactive ceramics*, Yamamuro T. L., Hench L., and Wilson J. (Eds.). Boca Raton, FL: CRC Press, 133–142.
27. Ji H. and Marquis P. M. 1993. Effect of heat treatment on the microstructure of plasma-sprayed hydroxyapatite coating. *Biomaterials*, 14: 64–68.
28. Filiaggi M. J., Coombs N. A., and Pilliar R. M. 1991. Characterization of the interface in the plasma-sprayed HA coating/Ti-6Al-4V implant system. *Journal of Biomedical Materials Research*, 25: 1211–1229.
29. Wang B. C., Chang E., and Yang C. Y. 1994. Characterization of plasma-sprayed bioactive hydroxyapatite coatings—in vivo and in vitro. *Materials Chemistry and Physics*, 37: 55–63.
30. Yang Y. C. and Chang E. 2003. The bonding of plasma-sprayed hydroxyapatite coatings to titanium: effect of processing, porosity and residual stress. *Thin Solid Films*, 444: 260–275.
31. Cao Y., Weng J., Chen J., Feng J., Yang Z., and Zhang X. 1996. Water vapourtreated hydroxyapatite coatings after plasma spraying and their characteristics. *Biomaterials*, 17: 419–424.
32. Wolke J. G. C., Van der Waerden J. P. C. M., de Groot K., and Jansen J. A. 1997. Stability of radiofrequency magnetron sputtered calcium phosphate coatings under cyclically loaded conditions. *Biomaterials*, 18: 483–488.
33. Yang C. Y., Wang C., Chang E., and Wu J. D. 1995. The influences of plasma spraying parameters on the characteristics of hydroxyapatite coatings: a quantitative study. *Journal of Materials Science: Materials in Medicine*, 6: 249–257.
34. Cheng K., Zhang S., Weng W., Khor K. A., Miao S., and Wang Y. 2008. The adhesion strength and residual stress of colloidal-sol gel derived [beta]-tricalciumphosphate/fluoridated-hydroxyapatite biphasic coatings. *Thin Solid Films*, 516: 3251–3255.
35. Gu Y. W., Khor K. A., and Cheang P. 2003. In vitro studies of plasma-sprayed hydroxyapatite/Ti-6Al-4V composite coatings in simulated body fluid (SBF). *Biomaterials*, 24: 1603–1611.

36. Khor K. A., Gu Y. W., Pan D., and Cheang P. 2004. Microstructure and mechanical properties of plasma sprayed HA/YSZ/Ti-6Al-4V composite coatings. *Biomaterials*, 25: 4009–4017.
37. Dey A., Mukhopadhyay A. K., Gangadharan S., Sinha M. K., and Basu D. 2009. Characterization of microplasma sprayed hydroxyapatite coating. *Journal of Thermal Spray Technology*, 18: 578–592.
38. Khor K. A., Li H., and Cheang P. 2003. Characterization of the bone-like apatite precipitated on high velocity oxy-fuel (HVOF) sprayed calcium phosphate deposits. *Biomaterials*, 24: 769–775.
39. Nieh T. G., Jankowski A. F., and Koike J. 2001. Processing and characterization of hydroxyapatite coatings on titanium produced by magnetron sputtering. *Journal of Materials Research*, 16: 3238–3245.
40. Dey A., Mukhopadhyay A. K., Gangadharan S., Sinha M. K., and Basu D. 2009. Development of hydroxyapatite coating by microplasma spraying. *Materials and Manufacturing Processes*, 24: 1249–1258.
41. Mancini C. E., Berndt C. C., Sun L., and Kucuk A. 2001. Porosity determinations in thermally sprayed hydroxyapatite coatings. *Journal of Materials Science*, 36: 3891–3896.
42. Pawlowski L. 1995. *The science and engineering of thermal spray coatings*. Chichester: John Wiley & Sons.
43. Geesink R. G. T., de Groot K., and Klein C. P. A. T. 1988. Bonding of bone to apatite-coated implants. *Journal of Bone and Joint Surgery*, 70: 17–22.
44. Dhert W. J. A., Verheyen C. C. P. M., Braak L. H., de Wijn J. R., Klein C. P. A. T., de Groot K., and Rozing P. M. 1992. A finite element analysis of the push-out test: influence of test conditions. *Journal of Biomedical Materials Research Part A*, 26: 119–130.
45. Zhang C., Leng Y., and Chen J. 2001. In vitro mechanical integrity of hydroxiapatite coatings on Ti6Al4V implants under shear loading. *Journal of Biomedical Materials Research*, 56: 342–350.
46. Lynn A. K. and Du Quesnay D. L. 2002. Hydroxyapatite coated Ti6-Al4-V part 1: the effect of coating thickness on mechanical fatigue behaviour. *Biomaterials* 23: 1937–1946.
47. Lynn A. K. and Du Quesnay D. L. 2002. Hydroxyapatite coated Ti6-Al4-V part 2: the effect of post deposition heat treatment at lower temperatures. *Biomaterials*, 23: 1947–1953.
48. Gledhill C., Turner I. G., and Doyle C. 2001. In vitro fatigue behaviour of vacuum plasma and detonation gun sprayed hydroxyapatite coatings. *Biomaterials*, 22: 1233–1240.
49. Mukherjee D. P., Dorairaj N. R., Mills D. K., Graham D., and Krauser J. T. 2000. Fatigue properties of hydroxyapatite coated dental implants after exposure to a periodantal pathogen. *Journal of Biomedical Materials Research Part A*, 53: 467–474.

7

Micro/Nanomechanical Properties
of Hydroxyapatite Coating

7.1 Introduction

In Chapter 6 we discussed the bulk or macromechanical properties of
HAp-based coatings. Now, beside its bonding strength, shear strength, and
fatigue behavior, it is also important to know the behavior of HAp-based
coatings at microstructural length scale because any failure or disintegration of material would primarily initiate at its microstructural length scale.
Further, if we look into the actual application of HAp coatings, they would
face mainly surface contacts rather than bulk contacts. In other words, for
such bioapplication, surface contacts is expected to play a role more dominant than that of the bulk contacts. Further, nanoindentation is one of the
potential tools to evaluate mechanical behavior at the microstructural length
scale. Therefore, in this chapter, more importance has been given to understanding the nanoindentation process and how it is useful for measuring
nanomechanical properties like nanohardness and elastic modulus of HAp
coatings. Moreover, it will be shown that prior to the present effort by the
author and coworkers, the work reported on especially the micro/nano-level
characterizations of the surface mechanical properties for HAp and HAp
composite coatings is far from significant, presumably because of the dif-
ficulties associated with the evaluation of the aforesaid properties for an
extremely heterogeneous microstructure, such as that of the plasma sprayed
HAp coatings. It is also a matter of big concern for prosthetic applications
of the HAp coatings that the coating lifetime may be and indeed is often
limited by premature failure due to contact or local delamination-induced
fracture. That is why evaluation and investigation of mechanical properties
at the local microstructural length scale of HAp-based coatings is so important and pertinent for further development of this biomaterial of tremendous
scientific and technological importance. We start our journey by presenting
next the basic theory of nanoindentation.

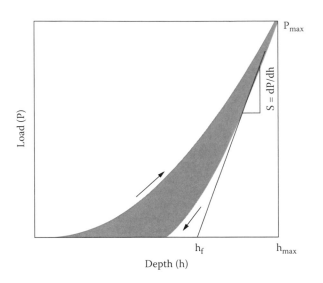

FIGURE 7.1
Schematic of typical load vs. depth of penetration (*P-h*) plot showing various parameters.

7.2 Basic Theory of Nanoindentation

During the nanoindentation, the high-resolution instrument continuously monitors the load, *P*, and depth of penetration, *h*, of an indenter, which in the present investigation was a Berkovich tip. These data are utilized to get the load vs. depth of penetration (*P-h*) data plot, schematically shown in Figure 7.1. The important physical quantities obtained from the load vs. depth of penetration plot are the peak load, P_{max}, maximum penetration depth, h_{max}, final penetration depth, h_f, and contact stiffness, *S*.

According to the Oliver-Pharr (O-P) model [1], which is the most commonly used method to obtain the hardness and Young's modulus of a material by instrumented nanoindentation, the nanohardness (*H*) is expressed as

$$H = P_{max}/A_{cr} \tag{7.1}$$

where P_{max} is the maximum applied load and A_{cr} is the real contact area between the indenter and the material. According to Oliver and Pharr, the polynomial form of A_{cr} can be expressed as [1]

$$A_{cr} = 24.56h_c^2 + C_1h_c + C_2h_c^{1/2} + C_3h_c^{1/4} + \ldots + C_8h_c^{1/128} \tag{7.2}$$

where C_1 to C_8 are constants to be determined by standard calibration method and h_c is the penetration depth determined from the following expression [2]:

$$h_c = h_{max} - \kappa(P_{max}/S) \tag{7.3}$$

where $\kappa \approx 0.75$ for a Berkovich indenter [3, 4].

Again, the contact stiffness (S), which is the slope of the first ~1/3 linear part recorded during the unloading cycle of the load vs. depth of penetration plot, can be expressed as [4]

$$S = (dP/dh)_{h=hmax} = \alpha C_A E_r \sqrt{A_{cr}} \tag{7.4}$$

where $\alpha = 1.034$ and $C_A = 2/\sqrt{\pi}$ for a Berkovich indenter [4], and E_r is the reduced Young's modulus. Following the O-P model, E_r can be expressed as [1]

$$1/E_r = (1 - v_i^2)/E_i + (1 - v_s^2)/E_s \tag{7.5}$$

where E and v are the Young's modulus and Poisson's ratio, respectively, and subscripts i and s denote the indenter and sample, respectively. For the Berkovich diamond indenter used in the present work, the values of E_i and v_i are taken as 1140 GPa and 0.07, respectively, following [4]. However, according to Oliver and Pharr, the unloading curve simply obeys the following power law [2]:

$$P = \alpha \, (h - h_f)^m \tag{7.6}$$

where α and m are empirical constants that can be determined by fitting the experimentally measured data from the load vs. depth of penetration data plot to Equation (7.6). Thus, the contact stiffness can also be determined using the following expression [2]:

$$S = (dP/dh)_{h \,=\, hmax} = \alpha m \, (h - h_f)^{m-1} \tag{7.7}$$

Therefore, substituting the values of S, α, C_A, and A_{cr} in Equation (7.5), the value of the reduced modulus E_r was calculated. The Young's modulus value of the sample, E_s, can then easily be obtained from Equation (7.5), using the known values of E_r and E_i.

7.3 Hardness

What is meant by hardness? Classically speaking, hardness is nothing but the surface resistance of any material to deform plastically. Now in the case of HAp coatings, higher hardness will ensure higher lifetime, particularly when there is a presence of micro-motion [5, 6], and cement-less fixation [6], where surface contact is a major factor, which is ultimately controlling the coating's long-term stability. For such biomedical applications, characterization of hardness at the microstructural level and examination of its anisotropy, if any, become issues of crucial importance because it is at this level that the relevant governing factors are decided. In addition, the anisotropy in hardness, if any, would also have very important implications in terms of the possible competition between compressive strain sharing in the plan and cross sections of such coatings, and thereby the possible extent of stress sharing between the coating and the metallic implant during the actual in-service applications. Such information will therefore also provide useful database and scientific understanding for the biomedical engineering community in terms of the development of the fail-safe design of bioactive HAp-coated metallic implants for various kinds of bone damage repairing applications.

7.3.1 What Does the Literature Say about Hardness?

A critical survey of existing knowledge base points out that in spite of the wealth of literature for hardness of HAp or HAp composite coatings (Table 7.1) [7–23], there has not been much of a systematic study on the nanohardness values or hardness (*H*) at the local microstructural level of HAp or HAp composite coatings on metallic substrates measured by the nanoindentation technique.

In the case of magnetron sputtered ultra-thin (~350–650 nm) films of HAp [9, 10] on Ti and Si substrates, nanohardness (*H*) data were a strongly sensitive function of film thickness and the type of substrate. The nanohardness of 6.5–8.5 GPa were reported across the depth range of 5–225 nm. Further, the HAp coatings on Ti/Si composites exhibited that nanohardness (H) could vary from about 10 to 4 GPa as the depth of penetration was increased from about 5 nm to 225 nm. This information implies that there can be a very strong indentation size effect present at the very small depth of the HAp coatings.

For the high-velocity oxy fuel (HVOF) sprayed HAp coatings on the Ti-6Al-4V substrates [11], the nanohardness data measured at the coating side were reported to be much higher than those evaluated at the coating-substrate interface. In a sharp contrast to this scenario, simply the opposite trend was reported for Nd-YAG laser-deposited HAp coatings on Ti substrates [12, 13].

The comparison of the data reported in [11–13] highlights the fact that the key factors that control the nanohardness data evaluated on the HAp coatings and at the interfaces between the HAp coatings and the respective substrates are yet to be fully understood.

Even for laser-deposited HAp coatings on Ti substrates, the nanohardness data varied to a large extent [12–14]. It was claimed that the nanohardness of the HAp–carbon nanotube (CNT) (5–20 vol%) composite was about 30% higher than that of the HAp matrix material [13]. Flame sprayed HAp coating

TABLE 7.1

Literature Status on Hardness of Pure HAp or HAp Composite Coating

Coating/Substrate: Processing Route	Surface of the Coating Studied: Technique of the Indentation	H (GPa)	Ref.
HAp/Ti-6Al-4V: MAPS	CS: Knoop microindentation	250–100 (H_k)	[7]
HAp/Ti-6Al-4V: MAPS	PS: Vicker's microindentation	4.3–2.7	[8]
HAp/Ti and HAp/Si: MS	PS: Berkovich nanoindentation	8.5–6.5	[9]
HAp/Ti and HAp/Ti-Si: MS	PS: Berkovich nanoindentation	10–4	[10]
HAp/Ti-6Al-4V: HVOF	CS: Nanoindentation	5.22–4.21	[11]
HAp/Ti: LD	CS: Vicker's nanoindentation	9–7	[12]
HAp-CNT/Ti-6Al-4V: LD	—: Berkovich nanoindentation	13.34–9.35	[13]
HAp/Ti and HAp/Si: LD	—: Nanoindentation	0.4–2.3	[14]
HAp/Ti: FS	PS: Berkovich nanoindentation	7	[15]
HAp/Ti-6Al-4V: MAPS	CS: Vicker's microindentation	380–300 (H_v)	[16]
HAp/Ti-6Al-4V: MAPS	CS: Knoop microindentation	310–210 (H_k)	[17]
HAp-YSZ/Ti-6Al-4V: MAPS	—: Vicker's microindentation	450–350 (H_v)	[18]
HAp/Ti-6Al-4V and HAp-Ti-6Al-4V/Ti-6Al-4V: MAPS	CS: Vicker's microindentation	4.4–3.1	[19]
HAp/Ti: FS	CS: Berkovich nanoindentation	8–4.5	[20]
HAp/SS316L: MIPS	PS: Berkovich nanoindentation CS: Berkovich nanoindentation	6–2.5 5–1.5	[21–23]

Note: CS = cross section, PS = plan section, MAPS = macroplasma spraying technique, MIPS = microplasma spraying technique, MS = magnetron sputtering technique, HVOF = high-velocity oxy fuel spraying technique, LD = laser-assisted deposition technique, FS = flame spraying technique, H_v = Vicker's hardness number, H_k= Knoop hardness number.

on the Ti substrate showed a great variation of nanohardness (~8–4.5 GPa) measured on both plan and cross sections [15, 20].

However, the basic reasons that contribute to such spectacular variations in nanohardness data of such HAp coatings, as mentioned herein, are yet to be fully understood, and thereby deserve thorough attention of the research community.

The composite coating of MAPS 50 vol% HAp/50 vol% Ti-6Al-4V on Ti-6Al-4V substrates showed a Vicker's hardness (H_v) of ~380 in the as-deposited condition. But it severely degraded, e.g., by about 20% to H_v of ~300, when as the deposited coating was immersed in simulated body fluid (SBF) up to 8 weeks [16]. On the other hand, it has been reported that the MAPS-HAp-YSZ/Ti-6Al-4V composite coatings showed H_v of 390–450 Kg.mm^{-2} when the as-deposited coatings were post-heat treated at 600–700°C for 3–12 h [18].

It is interesting to note that in the case of HVOF-HAp coatings of 100 μm thickness on Ti-6Al-4V substrates, nanoindentation experiments conducted at a constant load of 30 mN on the polished cross section gave an H value of ~5.22–4.21 GPa as one moved away from the coating side to the coating-substrate interface [11, 19]. These data suggested that the nanohardness was higher on the coating side than at the site of the coating-substrate interface.

However, the opposite trend was reported [12] for HAp coating prepared by laser surface engineering (Nd-YAG Laser) on Ti. These workers reported the nanohardness data evaluated at 500 μN load with a Vicker's diamond pyramidal indenter on the coating cross section [12]. The experiments were conducted at a loading rate of 50 μN.s^{-1}. It was reported that the coating had a nanohardness of 7 GPa, which was enhanced by about 30% to ~9 GPa at the site of the coating-substrate interface.

The microindentation technique was also utilized to evaluate hardness of MAPS-HAp [7, 8, 17, 18, 19] and MAPS-HAp composite [18, 19] coatings on Ti-6Al-4V substrates. The reported magnitudes of hardness had large variations, depending on the method of measurement adopted, e.g., a Vicker's [16] or a Knoop [7, 17] indentation.

The data given in Table 7.1 showed that most researchers (~60%) used the nanoindentation technique, while others (~40%) used the microindentation technique. Barring a few [20], the reported nanoindentation data were not systematic in terms of the study of load dependencies. Out of the 15 research publications, 47 and 33% measurements were on cross and plan sections, respectively, while three reports [13, 14, 18] were ambiguous about the locations where the measurements had been done.

From all the information presented above, it becomes apparent that in spite of the wealth of literature data, a general framework of understanding about the relative influences of several associated factors, e.g., microstructure, method of preparation, method and location of measurement in the coating, film thickness, relative depth of indentation, etc., on the hardness of HAp coatings is yet to emerge.

7.3.2 Nanohardness of MIPS-HAp Coatings

Recently, the nanohardness of MIPS-HAp coatings was evaluated by the present author and coworkers [21–23]. The inherent characteristic statistical variability of the data was treated through the application of Weibull statistics. The choice of such an unconventional approach is justified by the fact that characteristically the coating had a highly porous and heterogeneous microstructure. This aspect will be discussed in detail later.

They [21, 22] found that at a low load of 10 mN, the coating showed a nano-hardness value of ~5 GPa at a depth of about 170 nm, which dropped by 60%, e.g., ~2 GPa, at a depth of about 3000 nm for a higher load of 1000 mN. These data [21, 22] suggested the presence of a strong indentation size effect (ISE) in the nanohardness behavior of the coatings. Based on the experimental data and the scanning electron microscopy observations, it has been suggested by them [21, 22] that the extent of interaction of the indenter with average size defects across the depth of a MIPS-HAp coating could be responsible for the observed ISE [21, 22].

Further, at comparable loads, the nanohardness data of the MIPS-HAp coating measured on the cross section were usually higher than those measured on the plan section of the coating [21–23], thereby showing a characteristic anisotropy. As revealed by electron microscopy images and evaluation of the volume percent open porosity obtained from image analysis of the heterogeneous microstructure, this anisotropy in nanohardness was linked [23] to the presence of larger volume percentage of the open porosity, as well as the higher spatial density of planar defects, pores, and cracks in the plan section than in the cross section.

7.4 Young's Modulus

For prosthetic application of HAp-coated metallic implants, the Young's modulus (E) is the most important mechanical property because it provides the basis for the extent of strain sharing between the coating and the substrate. If anisotropy is present in Young's modulus of HAp coating, the amount of strain sharing will vary in different directions. Such a situation will certainly affect the mechanical integrity of the coating with respect to the direction of load application. Therefore, it is important to examine the presence or absence of anisotropy in Young's modulus of a given HAp coating.

7.4.1 What Does the Literature Say about Young's Modulus?

A critical survey of the relevant literature data (Table 7.2) indicates that in spite of the wealth of literature on the Young's modulus of HAp or HAp composite coatings [7–27], there has not been much of a systematic study on the evaluation of the Young's modulus at the scale of the local microstructural length scale level of HAp or HAp composite coatings deposited on metallic substrates and measured by the nanoindentation technique.

In the case of the magnetron sputtered HAp [9, 10] thin (~650 nm) films on Ti and Si substrates, the nanoindentation experiments conducted with

TABLE 7.2

Literature Status on Young's Modulus of Pure HAp or HAp Composite Coatings

Coating/Substrate: Processing Route	Surface of the Coating Studied: Technique of the Indentation	E (GPa)	Reference
HAp/Ti-6Al-4V: MAPS	CS: Knoop microindentation	52–12	[7]
HAp/Ti: MS HAp/Si: MS	PS: Berkovich nanoindentation	150–100	[9]
HAp/Ti: MS HAp/Ti-Si: MS	PS: Berkovich nanoindentation	150–70	[10]
HAp/Ti-6Al-4V: HVOF	CS: Nanoindentation	123–84	[11]
HAp/Ti: LD (Nd-YAG)	CS: Vicker's nanoindentation	160–80	[12]
HAp-CNT/Ti-6Al-4V: LD (Nd-YAG)	—: Berkovich nanoindentation	190–157	[13]
HAp/Ti: ArF-PLD HAp/Si: ArF-PLD	—: Nanoindentation	127–68	[14]
HAp/Ti: FS	PS: Berkovich nanoindentation	100	[15, 20]
HAp/Ti-6Al-4V: MAPS	CS: Knoop microindentation	37–47	[16]
HAp/Ti-6Al-4V: MAPS	CS: Knoop microindentation	23–27	[17]
HAp-YSZ/Ti-6Al-4V: MAPS	—: Knoop microindentation	65–50	[18]
HAp/Ti-6Al-4V: MAPS HAp-Ti-6Al-4V/Ti-6Al-4V: MAPS	PS, CS: Knoop microindentation	$(8–25)$-E_{ps} $(26–37)$-E_{cs}	[19]
HAp/SS316L: MIPS	PS: Berkovich nanoindentation CS: Berkovich nanoindentation	100–58	[21, 22]
HAp/Ti-6Al-4V: MAPS	CS: Berkovich nanoindentation	120–80	[24, 25]
HAp/Ti-6Al-4V: HVOF HAp-Ti/Ti-6Al-4V: HVOF	CS: Vicker's microindentation	77–32	[26]
HAp/Ti-6Al-4V: SG Fluoridated HAp/Ti-6Al-4V: SG	CS: Berkovich nanoindentation	47–74	[27]

Note: CS = cross section, PS = plan section, MAPS = macroplasma spraying technique, MIPS = microplasma spraying technique, MS = magnetron sputtering technique, HVOF = high-velocity oxy fuel spraying technique, LD = laser-assisted deposition technique, PLD = pulse laser deposition technique, FS = flame spraying technique, SG = sol-gel technique.

a Berkovich indenter gave a Young's modulus of about 100 GPa at a depth of about 5 nm on the plan section, but the Young's modulus increased by about 50% when the depth of penetration was increased to about 225 nm [9]. Further, it was reported that the HAp thin (~350–650 nm) films on Ti and Ti-Si composite substrates showed that Young's modulus was ~70–150 GPa [10].

This result suggested possibly that the measured value of Young's modulus could be sensitive to film thickness and also the type of substrate utilized for film deposition. These data also possibly highlight the need to do further work on how the measured values of the Young's modulus would be affected by the thickness of the deposited films and the chemical compositions of the substrates.

In the case of high-velocity oxy fuel (HVOF) sprayed [11, 26] HAp and HAp-Ti composite coatings on Ti-6Al-4V substrates, the Young's modulus measured by the Vicker's microindentation technique [26] on the cross section of the coating was reported to be ~77–32 GPa. However, Young's modulus measurement by the nanoindentation technique at a constant load of 30 mN on the cross section gave a much higher Young's modulus value of 123–84 GPa as one moved away from the coating side to the coating-substrate interface [11].

This information would strongly reiterate the fact that the Young's modulus values could be sensitive to the method of measurement and the location, e.g., at the coating or at the coating-substrate interface, where the measurements are made. It also suggests that the Young's modulus of the HAp coatings may be higher at the coating side than at the coating-substrate interface site, where it can register a substantially lower magnitude [11]. It also frames another query: What is the reason behind the drop in the magnitude of the Young's modulus at the coating-substrate interface?

In sharp contrast to these observations, the opposite nature was reported for HAp coating deposited by the Nd-YAG laser deposition technique [12, 13] on Ti substrates. In this particular case, the Young's modulus was measured by the nanoindentation technique with a Vicker's diamond pyramidal indenter under 500 μN load applied on the coating cross section [12]. It has been reported by these workers [12] that the Young's modulus was about 80 GPa on the coating side, but it increased by almost 100% to about 160 GPa at the coating-substrate interface. This information clearly demonstrated that the modulus could increase by as much as 100% as one moved away from the coating side to the coating-substrate interface.

On the other hand, the pulsed laser-deposited HAp coatings on the Ti and Si substrates showed a Young's modulus of ~127–68 GPa when measured by the nanoindentation technique [14]. A comparison of the results reported in [12–14] would possibly indicate that even for a given technique of film deposition (e.g., laser-induced deposition), a given substrate (e.g., Ti), and a given measurement technique (e.g., nanoindentation), the measured values of Young's modulus could vary to a considerably large extent. There is apparently no explanation available in literature for such a wide scatter in the data of the same property measured by a similar technique applied to HAp coatings deposited in similar synthesis processes on a given type of metallic (e.g., Ti) substrates.

The underlying scientific reasons that can cause such a wide variation of the Young's modulus data are therefore an issue that definitely deserves much more attention for future development of the HAp coatings for biomedical implant applications.

Attempts have also been made to increase the Young's modulus of HAp coating by reinforcing with CNT. For a HAp matrix reinforced with 5, 10, and 20 vol% of CNT [13], the Young's modulus was reported to be as high as ~160–190 GPa when measured by the nanoindentation technique with a

Berkovich indenter. It was claimed that the Young's modulus of the composite was about 2–17% higher than that of the HAp matrix material. It was very interesting to note that this report [13] has, until now, remained the singular yet a very important attempt to apply the CNTs for improving the nano-mechanical properties of the HAp coatings.

In this connection, it will be very pertinent to identify the fact that from the survey of existing literature data, as discussed in the present work, it emerges as a very important point of focus in future research that a lot of work needs to be done to examine the relative efficacies of the new carbon materials as the reinforcing agents that can substantially improve the mechanical properties of the HAp coatings. The examples of such new carbon materials may include single-wall carbon nanotube (SWCNT), multiwall carbon nanotube (MWCNT), nanocarbon powders, nanosize carbon/graphite flakes, diamond-like carbons (DLCs), micron-size and nanosize carbon fibers, graphene sheets or grapheme oxides, etc., especially for the load, as well as friction-bearing prosthetic applications, since many of these new carbon materials are already proven to be nicely biocompatible. At present, the situation is so poor that even the very preliminary database does not yet exist.

In the case of flame sprayed HAp coating on Ti substrate, Young's modulus of 100 GPa was measured on the plan section of the coating by the nanoindentation technique with a Berkovich indenter [15, 20]. However, a comparatively much lower Young's modulus of ~47–74 GPa was measured by the nanoindentation technique with a Berkovich indenter for the cross sections of sol-gel-derived HAp and fluoridated-HAp coatings [27].

A comparison of the results reported in [15, 20, 21] would suggest, again, that even for a given technique, e.g., nanoindentation, and a given indenter, e.g., the Berkovich indenter, the Young's modulus of HAp coatings could be a sensitive function of the preparation technique employed and the resultant microstructure produced. What is not yet known is what factors can effectively contribute to such situations. Such issues therefore again assume a very pertinent and important role in future research on the HAp coatings as a bioactive ceramic material.

On the other hand, the Young's modulus measured at a constant load of 3 mN by the nanoindentation technique on the cross section of MAPS-HAp coating on Ti-6Al-4V substrate showed magnitudes of 120 GPa in the crystalline zone and only about 80 GPa in the amorphous zone [24, 25]. This information again suggested that the magnitude of Young's modulus was sensitive to the location of data evaluation, as was also noted earlier.

However, the cross section of MAPS-HAp coatings deposited on the Ti-6Al-4V substrates showed much lower Young's moduli values of ~52–12 GPa [7], 47–37 GPa [16], and 27–23 GPa [17] when measured using the microindentation technique with a Knoop indenter. Similarly, the MAPS-HAp and functionally graded HAp-Ti-6Al-4V composite coatings showed the Young's moduli of ~8–25 GPa and ~26–37 GPa for both plan and cross sections, respectively [19], when measured by the microindentation technique using a Knoop indenter.

A comparison of the results reported for MAPS-HAp coatings in [7, 16, 17, 19, 24, 25] again further emphasizes that even for a given type of coating, e.g., MAPS-HAp on a given substrate, e.g., Ti-6Al-4V, depending on the method of measurement, the magnitude of Young's modulus may vary to a very significant extent. In addition, it should be mentioned that when measured by a technique similar to that in [19], a comparatively higher Young's modulus of ~65–50 GPa was again reported for the macroplasma sprayed composite coatings of HAp and yittria-stabilized zirconia [18].

Of the reported data in the literature (Table 7.2), about 65% used the nanoindentation technique, while the other 35% used the microindentation technique, thereby suggesting that the emerging nanoindentation technique is a viable and important method for the measurement of Young's modulus in HAp and HAp composite coatings. Most of the reported data were not systematic in terms of load dependence study, barring a few, e.g., [20]. Further, out of the 17 reported results, 59% of the experiments were conducted on the cross section and only about 29% were conducted on the plan section, while three reports [13, 14, 18] did not clearly mention unequivocally whether the measurement was done on the plan section or on the cross section of the HAp or the HAp composite coating.

Further, the Young's modulus was reported to be sensitive to several factors. These include the relative depth of indentation with respect to film thickness, e.g., the presence or absence of a substrate mechanical property effect on the measured value [28], and also the type of substrate utilized for film deposition [9, 10], the method and the location of measurement, e.g., the crystalline zone or the amorphous zone [24, 25], at the coating, evaluating along the cross section from the coating side toward the coating-substrate interface [11, 19, 20, 26] or from the coating-substrate interface side toward the coating side [12–14], the presence or absence of a reinforcing phase [12, 18], the preparation technique employed, e.g., sol-gel or flame sprayed for the coating, the resultant microstructures produced and porosity [15, 20, 27], etc.

Even for the similar laser-deposited HAp coatings on Ti substrate, Young's modulus measured by the nanoindentation technique varied to a large extent [12–14]. Similarly, depending on the method of measurement for MAPS-HAp coatings on Ti-6Al-4V substrate, there was large variation in the Young's moduli data [7, 16, 17, 19, 24, 25]. The real scientific reasons that give rise to such large variations in the Young's modulus are yet to be understood, and therefore should be considered a major issue for further research to develop better HAp coatings for future application.

Thus, in spite of the numerous data reported in literature, a focused analysis of the elastic behavior of the HAp coatings is still missing. It appears that from the multitude of data reported in literature, a general structure of understanding about the relative influences of several associated factors, e.g., microstructure, method of preparation, porosity, method of deposition and location of measurement, residual stress, coating or film thickness, relative depth of indentation, and the influence of a substrate's mechanical properties, on the Young's modulus of HAp coatings is yet to come into sight.

7.4.2 Young's Modulus of MIPS-HAp Coatings

The range of Young's moduli data were reported to be about 100 to 60 GPa at a wide range of indentation load, i.e., 10 to 1000 mN loads for the MIPS-HAp coatings developed by Dey and coworkers [21, 22]. Although a direct comparison of the data of present work with literature data was not possible because the processing method and measurement methods were not all identical, it may still be mentioned that the data of the present work compared favorably with literature data (Table 7.2).

However, the data reported [12] for a laser-deposited HAp coating was slightly higher than those measured in the work reported by Dey et al. for MIPS-HAp coatings [21, 22]. We know that the Young's modulus value of sintered bulk HAp is reported as 125 GPa [29]. However, the average Young's modulus of the MIPS-HAp coatings was measured [21, 22] to be 80 GPa, which is in fact much lower than that of the bulk sintered HAp material. This was due to the fact that the coating had a high porosity of about 18–20 vol%.

Work done on plasma sprayed ceramic thermal barrier coatings has already established that volumetric and planner defects could cause significant reduction of Young's modulus values [30, 31]. The dependence of Young's modulus (E) on vol% porosity (p) is expressed as [32]

$$E = E_0 \exp (-bp) \tag{7.8}$$

where b is a constant (~2.5), and E_0 is the Young's modulus for a material with zero porosity, i.e., of theoretical density. Assuming $E_0 = 125$ GPa [29], for a HAp coating with 18 vol% porosity, the predicted Young's modulus value was ~76 GPa, which matched quite well with the load-averaged Young's modulus data of 80 GPa, as mentioned above.

7.5 Effect of SBF Immersion

The HAp coatings prepared on Ti-6Al-4V and Ti-13Nb-11Zr alloys by the biomimetic route showed the hardness values of H_v ~ 48.5–47 and ~ 54–58, respectively, as the coated sample was exposed in vitro for 4–12 weeks in SBF solution [28]. The experiments were conducted on the plan sections, i.e., on the top surfaces of the coatings, by an automated Vicker's microhardness tester at a load of 0.05 kgf with 200 μm as the maximum depth of penetration.

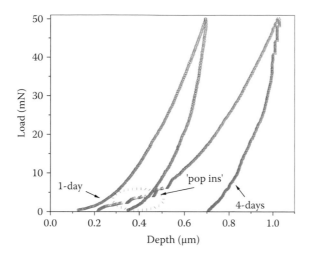

FIGURE 7.2
P-h plots of the MIPS-HAp coating after immersion in the SBF solution.

7.5.1 Effect of SBF Immersion for MIPS-HAp Coatings

The present authors have studied both nanohardness and elastic modulus of MIPS-HAp coatings following immersion in SBF. The nanohardness of the as-sprayed MIPS-HAp coating decreased from ~4.5 GPa to ~3 GPa after 1 day immersion in the SBF solution [33]. This occurred as the dissolution process was dominant after just 1 day of immersion. It was further reported [33] that after 1 day of immersion in the SBF solution, the nanohardness again increased to ~3.5 GPa. This happened as the apatite layer deposition process was dominant during the subsequent 4 to 7 days of immersion in the SBF solution. Typical load-depth plots of the MIPS-HAp coating after SBF immersion are shown in Figure 7.2 for 1 and 4 days, respectively. Multiple "pop-ins" shown in the load-depth (*P-h*) plot (Figure 7.2) for 4 days of immersion in SBF had occurred due to the evolution of the nanoporous apatite structure formation. However, the nanohardness after 14 days of immersion decreased only marginally, if at all, in comparison to the nanohardness measured at 4 and 7 days of immersion in the SBF solution. The elastic modulus data as a function of immersion time also followed a trend (e.g., from ~100 to ~60 GPa) similar to what was observed in the nanohardness data [33].

7.6 Reliability Issues in Nanoindentation Data

Several researchers, including the present authors [21–23, 33], performed numerous indentation arrays at each load at various locations of PS-HAp coatings chosen without any bias, as the nature of the coating was porous and heterogeneous. In general, the scatter in data was very high for the plasma sprayed coatings, presumably due to the highly heterogeneous and porous microstructure of the coatings.

The complex microstructure of the present coating consisted of micropores, macropores, intersplat and intrasplat cracks, and other unavoidable defect features, as depicted earlier in Chapter 4. This characteristic heterogeneity of the microstructure is reflected in the characteristic scatter present in the experimentally measured data on nanohardness and Young's modulus. To treat this scatter, a two-parameter Weibull distribution function was utilized and the corresponding characteristic values in each case were evaluated to obtain suitable results for structural designing purposes. Before proceeding further, it thus becomes imperative to provide just the basic idea of the Weibull distribution function in the subsection that follows.

7.6.1 Weibull Distribution Function

The Weibull distribution function provides the probability, p, for a given parameter, x, as

$$p = 1 - \exp\left[-(x/x_o)^m\right] \tag{7.9}$$

where x_o is a scale parameter whose probability of occurrence is 63.2%, and m is known as the Weibull modulus. m is a dimensionless quantity and indicates the degree of spread or scatter in the experimentally measured data of x. A larger magnitude of m implies that the degree of scatter in the experimentally measured data is small. The opposite behavior, i.e., a small magnitude of m, signifies the degree of scatter in the experimentally measured data is large. Further, the survival probability of the ith observation in the data arranged in ascending order can be written as

$$p = (i - 0.5)/N \tag{7.10}$$

where N is the total number of observations. Although, in some literature, the formula for the probability estimator p was found to be different, here we have employed the most extensively used expression. Taking ln for two times both the sides and simplifying, Equation (7.10) can be expressed as

$$\ln[\ln\{1/(1-p)\}] = m[\ln(x) - \ln(x_o)] \tag{7.11}$$

The values of m and x_o can be obtained by fitting (least-squares regression) the experimental data to Equation (7.11). The slope of the straight line will provide the value of m, and the intercept on the Y axis will give the value of x_o. Finally, by setting the value of $\ln[\ln\{1/(1-p)\}]$ equal to zero and placing the value of m in Equation (7.11), we can easily compute the characteristic value

(x_o) of the related parameter, x. The characteristic values (x_o) are of immense engineering importance, as they provide the designer with a unique and dependable value of the required parameter. In the present work described here, x is nanohardness and Young's modulus. The corresponding characteristic values are known as H_{char} and E_{char}.

7.6.2 Weibull Modulus of Nanohardness and Young's Modulus of MIPS-HAp Coating

The Weibull distribution fittings for the nanohardness data of the MIPS-HAp coatings are displayed for the low (10–100 mN) loads in Figure 7.3a and for the high (300–1000 mN) loads in Figure 7.3b. Similarly, the Weibull distribution fittings for the Young's modulus of the coating, as determined by the nanoindentation experiment, are displayed for the low (10–100 mN) loads in Figure 7.3c and for the high (300–1000 mN) loads in Figure 7.3d. The m values of the nanohardness data varied in the range 2–8. The m value increased steadily beyond 100 mN, with the load value close to 8 at a higher load of

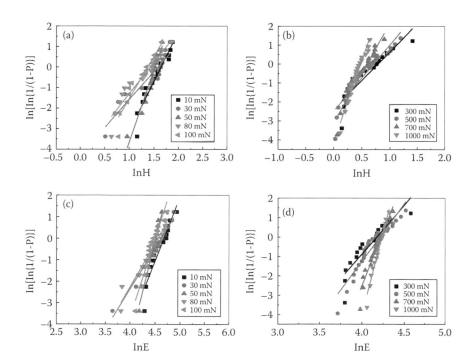

FIGURE 7.3 (See color insert.)
Weibull distribution plots of nanohardness at (a) low (10–100 mN) and (b) high (300–1000 mN) loads. Weibull distribution plots of Young's modulus at (c) low (10–100 mN) and (d) high (300–1000 mN) loads.

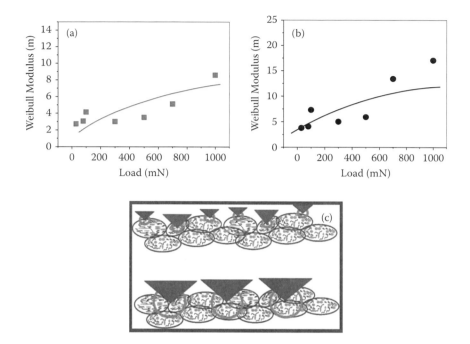

FIGURE 7.4 (See color insert.)
Weibull modulus of (a) nanohardness and (b) Young's modulus as a function of load of the MIPS-HAp coatings. (c) Model of indenter-MIPS-HAp coating interaction. (Reprinted/modified from Dey et al., *Ceramics International*, 35: 2295–2304, 2009. With permission from Elsevier. Reprinted/modified from Dey et al., *Journal of Materials Science*, 44: 4911–4918, 2009. With permission from Springer.)

1000 mN (Figure 7.4a). A similar trend was exhibited by the *m* values of the Young's modulus data (Figure 7.4b).

Here we propose an explanation for such observation. The physical picture of the aforesaid phenomenon is schematically shown in Figure 7.4c, and it is proposed [21, 22] to be linked to the length scale of interaction phenomena between the penetrating nanoindenter and the characteristic microstructural scale flaws and defects of the MIPS-HAp coating.

The situation is like this that when the magnitude of the applied load is in the lower side (e.g., $P \leq 300$ mN), the nanoindentation size scaled with the size of microstructural defects, which is drawn schematically in Figure 7.4c (upper part), and it may negotiate many intersplat boundaries (i.e., the typical weak and defect-populated zones in the microstructure). This circumstance perhaps produces a high degree of interaction, which finally caused a relatively higher degree of scatter in both nanohardness and Young's modulus data. This picture corroborates well with the relatively lower *m* value, as had been indeed observed in the experimental data of the present authors [21, 22]. However, an opposite picture had emerged during a nanoindentation study conducted under a higher load regime (e.g., 300 mN $< P \leq 1000$ mN) on the MIPS-HAp coatings. Here, the nanoindentation size was much larger. So, it covered a much larger number of splats. Consequently, indents had the chance to negotiate with only a small number of intersplat boundaries. This situation is schematically shown in the lower part of Figure 7.4c. Therefore, relatively less chances of the interaction between the nanoindenter and the characteristic microstructural scale flaws and defects are reflected in relatively lower scatter in the experimentally measured data [21, 22].

7.6.3 ISE in MIPS-HAp Coating

The variations of the characteristic values of nanohardness (H_{char}) and Young's modulus (E_{char}) are plotted in Figure 7.5a and b, respectively, as a function of load. H_{char} decreased from about 5 to 1.5 GPa (Figure 7.5a), and E_{char} decreased from about 100 to 63 GPa (Figure 7.5b), as the nanoindentation load was increased from 10 to 1000 mN. The nanohardness data apparently suggest the presence of a strong ISE, i.e., load dependency of the data.

Nevertheless, ISE is a much more complex problem than what is reflected in a simplistic fashion in the data, and that is why it is preferable to term it as *apparent ISE*. What happens in apparent ISE is that along with the original ISE, the influence of microstructural defect (i.e., characteristic microstructural flaws) would also be a factor toward the contribution of ISE. That's why Dey and his coworkers termed this an apparent ISE. The present authors and coworkers suggested [21, 22] that at low load, the extent of scatter in the nanohardness and Young's modulus data measured by nanoindentation was governed by the presence of very shallow surface and subsurface defects in the MIPS-HAp coating. However, those measured at relatively higher load of nanoindentation might be governed not only by the statistical distribution

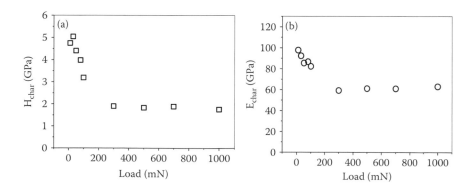

FIGURE 7.5

Characteristic values of (a) nanohardness (H_{char}) and (b) Young's modulus (E_{char}) as a function of load in MIPS-HAp coatings. (Reprinted/modified from Dey et al., *Ceramics International*, 35: 2295–2304, 2009. With permission from Elsevier. Reprinted/modified from Dey et al., *Journal of Materials Science*, 44: 4911–4918, 2009. With permission from Springer.)

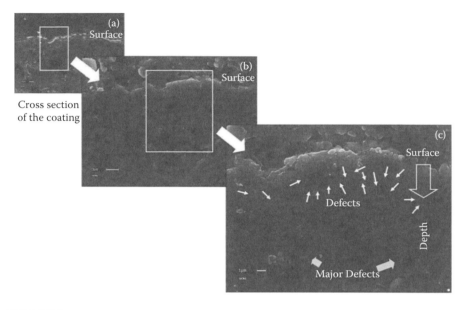

FIGURE 7.6
SEM photomicrographs of the cross section of the MIPS-HAp coating at (a) ×1000, (b) ×6000, and (c) ×10,000 magnifications. (Reprinted/modified from Dey et al., *Ceramics International*, 35: 2295–2304, 2009. With permission from Elsevier.)

of larger pores and deeper crack-like defects, but also by differences in their local orientations, which might have been conducive enough to ultimately provide an averaging out effect, so much as to reduce the overall range of the scatter in the corresponding data being evaluated.

The scanning electron microscopy (SEM)-based evidence taken from the coating cross section at progressively higher magnification showed [21, 22] that there was a gradual increase of fine micropores and microcracks, as well as macroscopic defects, such as large macropores and deeper cracks, as one traverses from the surface toward the depth of the coating. This situation is pictorially represented in Figure 7.6a–c, where SEM micrographs had been taken at progressively higher magnifications.

7.6.4 Anisotropy in MIPS-HAp Coating

The load-depth (*P-h*) plots for nanoindentation experiments conducted on both plan and cross sections of the coatings are shown in Figure 7.7a [23]. The *P-h* plot for the cross section of the MIPS-HAp coatings is much steeper than the *P-h* plot for the plan section. These data indicate that the higher magnitude of nanohardness should be at the cross section and not the plan section. It is indeed interesting to note that the data presented in Figure 7.7b and c reflect the presence of anisotropy in both nanohardness [23] and Young's modulus.

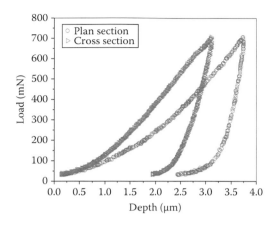

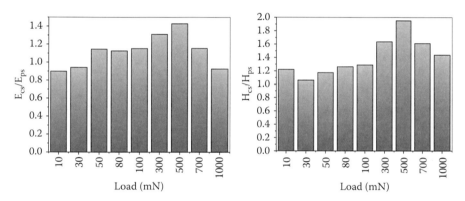

FIGURE 7.7 (See color insert.)

(a) Load-depth plots from nanoindentation experiments on MIPS-HAp coating at 700 mN on plan section and cross section. Ratio of (b) nanohardness data of the cross section (H_{cs}) and plan section (H_{ps}) as a function of load (replotted from Dey and Mukhopadhyay, *Advances in Applied Ceramics*, 109: 346–354, 2010) and (c) Young's modulus data of the cross section (E_{cs}) and Young's modulus data of the plan section (E_{ps}) as a function of load.

At any given load, the projected area of contact was always higher on the plan section than on the cross section, which resulted in higher magnitudes of both nanohardness and modulus of cross section over those for the plan section. The factors that could have contributed to such anisotropy could be many. These may typically include, but are not necessarily limited to, the spraying method, spraying condition, post-treatment, porosity, processing-induced natural crack population, shape and size of defects, etc. It will be pertinent here to recall the detailed field emission scanning electron microscopy (FESEM) studies presented in Chapter 4 in relation to the microstructure and inherent anisotropy of the same in the case of MIPS-HAp coatings. The higher magnitudes of both nanohardness and Young's modulus of the cross section, which lead to the observed anisotropy, were most likely linked to the presence of less porosity and microstructural defects in the cross section than in the plan section of MIPS-HAp coatings.

7.7 Fracture Toughness of MIPS-HAp Coatings

7.7.1 Why Is Fracture Toughness Important and How Is It Measured?

For orthopedic applications, the in-service lifetime is of utmost importance. The intrinsic resistance of a given ceramic against catastrophic crack propagation is called fracture toughness. As the in-service lifetime will be determined by the resistance of the coating against the propagation of a through-thickness crack, it is essential to know the fracture toughness behavior of such calcium phosphate- or HAp-based bioceramic coatings. Thus, it becomes undoubtedly most necessary to measure the plane strain fracture toughness (K_{1c}) at the scale of the microstructure because it is at this particular length scale where the intrinsic crack growth resistance of the HAp coating will control its ultimate structural integrity.

One of the most popular methods of toughness measurement is the indentation fracture method (IFM), where a crack is grown in a controlled manner on the surface of a given material. One reason for its popularity is that it is possibly the only method for measuring fracture toughness when a very small specimen volume is available for testing purposes. Now, in the IFM technique, either a Vicker's tip for microindentation or a Berkovich tip for nanoindentation is used for controlled crack growth. After that, the crack length should be measured using either the optical or electron microscopy technique. The following equation is most popularly used for calculation of the micro/nanoscale fracture toughness of HAp coating [34–36]:

$$K_{1c} = \alpha \left(\frac{E}{H} \right)^{0.5} \left(\frac{P}{C^{1.5}} \right) \tag{7.12}$$

where P is the applied load in nanoindentation, C is the crack length measured from the center of the nanoindent, and E and H are the Young's modulus and nanohardness of the coating, respectively. Generally, the value of the constant α is taken as 0.016 [36]. However, for different indenter geometries there can be slight variations in the value of α. Further, the nanoindentation technique has been identified to be superior than the microindentation technique because the crack length in the nanoindentation could be varied from a length scale less than the splat size up to a length scale that is compatible to the splat size in the HAp coating. Thus, the greatest advantage of this

TABLE 7.3

Literature Status on Fracture Toughness of Pure HAp or HAp Composite Coatings

Coating/Substrate	Method of Coating	Details of Experiment	K_{1c} (MPa•m$^{0.5}$)[a]	Reference
HAp/Ti-6Al-4V	MAPS	P = 0.3–0.5 kgf	0.63–1.14	[7]
HAp/Ti-6Al-4V	MAPS	P = 300 g	0.94–1.14	[19]
HAp-Ti-6Al-4V/Ti-6Al-4V	MAPS	P = 300 g	0.5–2.1	[19]
HAp/Ti-6Al-4V	HVOF	P = 0.5 N	0.49–0.54	[26]
HAp-TiO$_2$/Ti-6Al-4V	HVOF	P = 0.5 N	0.6–0.67	[26]
HAp/Ti-6Al-4V	MAPS	P = 100 g	0.39	[37]
HAp-4 wt% CNT/Ti-6Al-4V	MAPS	P = 100 g	0.61	[37]
HAp/Ti-6Al-4V	HVOF	P = 0.5 N	0.48	[38]
HAp + TiO$_2$/Ti-6Al-4V	HVOF	P = 0.5 N	0.6–0.67	[38]
HAp-YSZ-10–50%/Ti-6Al-4V	MAPS	—	0.92–1.5	[39]
HAp/Ti-6Al-4V	MAPS	—	0.55	[40]
HAp-YSZ/Ti-6Al-4V	MAPS	—	1.47–1.63	[40]
HAp/Ti-6Al-4V	MAPS	—	0.55	[41]
HAp + YSZ/Ti-6Al-4V	MAPS	—	0.84–1.09	[41]
HAp/Ti-6Al-4V	MAPS	—	0.46–0.95	[42]
HAp-TiO$_2$/Ti-6Al-4V	MAPS	—	0.47–1.76	[42]
HAp + α-TCP/Ti-6Al-4V	MAPS	—	1.05–1.41	[42]
HAp/SS316L	MIPS	100–1000 mN	0.6–0.4[b]	[43]

Note: MAPS = macroplasma spraying technique, MIPS = microplasma spraying technique, HVOF = high-velocity oxy fuel spraying technique, P = indentation load for the measurement of fracture toughness.

[a] Fracture toughness measured by Vicker's microindentation method.
[b] Fracture toughness measured by nanoindentation method.

technique is that it can evaluate fracture toughness value at the scale of the local microstructure. It is, however, appreciable that if indentation experiments are performed with Vicker's microindentation, then the spatial zone of indentation will surely be much larger than the splat size of plasma sprayed-HAp coatings, and in correspondence, the crack length will be much longer than the typical spatial extent of a splat. Since the indent size is much larger in microindentation, it is almost impossible to guarantee that the indents would be site specific and in regions populated with least defect. On the other hand, the most unique advantage of the nanoindentation technique is that it is possible to make site-specific indentations in regions where the coating is most dense and has the least amount of defects. A list of pertinent fracture toughness data is summarized in Table 7.3 [1, 26, 37–43].

7.7.2 Site-Specific Nanoindentation

When we have to measure the crack lengths from the footprints of nanoindentation, we have to be very careful regarding the interaction of the crack with

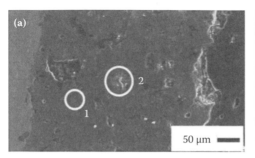
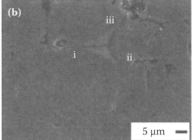

FIGURE 7.8
The example of site-specific indentation: SEM photomicrographs of the nanoindents on MIPS-HAp coatings at (a) 300 mN (marked as 1) and 700 mN (marked as 2) loads and (b) 300 mN load at higher magnification showing well-developed crack at three corners (marked as i, ii, and iii). (Reprinted/modified from Dey and Mukhopadhyay, *International Journal of Applied Ceramic Technology*, 8: 572–590, 2011. With permission from Wiley.)

microcracks that might preexist in the microstructure. For a plasma sprayed coating, this situation is very common. It will be shown in the following discussion how site-specific indentation can be intelligently used to rule out the aforesaid confusion. The present authors have shown [43] how to make site-specific nanoindentations in the most apparently dense area of a given porous MIPS-HAp coating. SEM micrographs of two nanoindents made at 300 mN (marked as 1, Figure 7.8a) and 700 mN (marked as 2, Figure 7.8a) are shown here as a typical example. It may be noted that the nanoindent made at 700 mN had severe interaction with preexisting microstructural defects, like pores and microcracks. Such data should therefore be excluded from the data pool for measurement of nanoindentation crack length at this particular load. In contrast, there was no defect interaction in the case of the indent made at the lower load of 300 mN (marked as 1 in Figure 7.8a and shown at higher magnification in Figure 7.8b). Thus, such data can be taken as a valid indent for measurement of nanoindentation crack length.

7.7.3 What Does the Literature Say?

A critical survey of pertinent literature data, as depicted in Table 7.3 [1, 26, 37–43], reveals that most of the K_{1c} data (average ~ 0.5 MPa•m$^{0.5}$) reported are evaluated only by Vicker's microindentation across the cross section of dense MAPS and HVOF sprayed HAp coatings on Ti-6Al-4V substrates. Considering that the bulk sintered polycrystalline HAp shows a K_{1c} value of about 0.7 to 1 MPa•m$^{0.5}$ [44, 45], it is not surprising that still higher K_{1c} values for even dense HAp coatings are not so frequently reported [19]. Reported K_{1c} data did not improve much even when a much finer HAp particle size was used [42], and it also decreased with load [7]. Further, several attempts were also made to enhance the fracture toughness of HAp coatings by the incorporation of a second phase, e.g., yttria-stabilized zirconia (YSZ) [42, 44, 45],

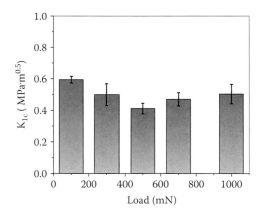

FIGURE 7.9
Load-dependent fracture toughness of MIPS-HAp coating. (Reprinted from Dey and Mukhopadhyay, *International Journal of Applied Ceramic Technology*, 8: 572–590, 2011. With permission from Wiley.)

titanium dioxide [41, 42], CNT [37], and fluorine [27]. It appears from the review of literature (Table 7.3) that the data in literature pertain to more of a global K_{1c} value, as the crack sizes generated by the loads chosen are much larger than the typical splat size. Moreover, systematic studies in terms of load variation and crack front interaction with the local splat microstructure were not present in the current literature of MAPS-HVOF HAp coatings.

However, Dey and Mukhopadhyay [43] have reported fracture toughness data of porous MIPS-HAp coatings measured for the first time ever by the nanoindentation technique. They [43] showed a very slightly decreasing trend of K_{1c} with load (Figure 7.9). The values of K_{1c} indicated that, barring a few [7, 19, 42], the present MIPS-HAp coating had a toughness value at least as high and as good as those reported for the dense thermally spread pure HAp (i.e., except addition of second phase) coatings. However, data of still higher magnitude, i.e., around 1 MPa•m$^{0.5}$ or more, may not be a realistic estimation of the coating toughness alone since dense sintered pure HAp showed toughness values in the range of 0.7 to 1 MPa•m$^{0.5}$.

7.7.4 Why Do MIPS-HAp Coatings Show High Toughness?

In this context, an extensive study of the cracking behavior around the nanoindents has been done through scanning electron microscopy by the present authors [43]. SEM micrographs corresponding to Berkovich nanoindent footprints of MIPS-HAp coatings are shown in Figure 7.10a and b for low load (100 mN) and high load (1000 mN), respectively. It can be seen that the radial cracks are well defined from near the three corners of any given Berkovich indent. However, sometimes the presence of micro- and macropores and micro- and macrocracks in the vicinity has a strong influence on

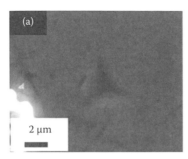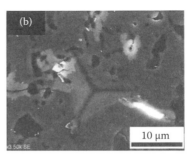

FIGURE 7.10

SEM photomicrographs of the Berkovich nanoindent impressions on the MIPS-HAp coating at low and high loads, e.g., (a) 100 mN and (b) 1000 mN. (Reprinted/modified from Dey and Mukhopadhyay, *International Journal of Applied Ceramic Technology*, 8: 572–590, 2011. With permission from Wiley.)

the cracking process around the nanoindent (Figure 7.10b). They [43] opined that high fracture toughness of MIPS-HAp coatings is linked with energy dissipation at the scale of microstructure by means of crack blunting, crack brunching, bifurcation, etc., and thus may contribute toward toughness enhancement of the MIPS-HAp coating.

Usually, fracture toughness is induced by two varieties of mechanisms. The first one is intrinsic mechanisms that function ahead of the crack tip. This offers materials inherent resistance to microstructural damage and cracking. However, this aforesaid mechanism is more predominant in metals (i.e., ductile materials), where it acts to increase resistance to crack initiation. In contrast, in a brittle ceramic coating like the present MIPS-HAp coatings under discussion, the second type of mechanism (i.e., the extrinsic mechanisms) may be predominant. This type of mechanism usually acts through crack bridging or microcracking [46, 47], as was also experimentally observed (Figure 7.10b) in the present MIPS-HAp coatings [43].

It has been proposed [48] that such extrinsic toughening may happen due to the development of a *frontal process zone* ahead of the growing crack and the resulting formation of a microcracking zone in the crack wake. Further, the resulting dilation and reduction in modulus that occur within this zone, if constrained by surrounding material, can act to shield the crack tip, and therefore extrinsically toughen the material [48]. It has been revealed by the present authors [43] that MIPS-HAp showed high toughness through the following possible modes:

1. Crack branching
2. Crack blunting at micropore and macropore
3. Crack deflection/bifurcation
4. Localized secondary/multiple cracking
5. Partial local delamination

It has been argued [43] that such energy dissipative processes, as mentioned above, can act on their own in a concerted manner, yet remain independent of each other. It is also possible to have a situation when all of the aforesaid processes can be simultaneously active to dissipate energy from the loading train. It is most likely then that through these processes of energy dissipation, as mentioned above, a marginally higher toughness was achieved in the MIPS-HAp coating. Thus, the MIPS-HAp coating developed by Dey and Mukhopadhyay [43] showed promising toughness behavior, and therefore might exhibit better reliability for in vivo application.

7.8 Summary

This chapter describes the different aspect of the micro/nanomechanical properties, like nanohardness, elastic modulus, and fracture toughness of HAp coatings evaluated at the local microstructural length scale. More importance has been given to plasma sprayed HAp coatings in general, and to MIPS-HAp coatings in particular. The nanoindentation technique, which is one of the potential tools to characterize thin films and the coating's micro/nanomechanical properties at the local microstructural length scale, has been explained in detail. The literature scenario regarding nanohardness, Young's modulus, and fracture toughness of HAp coatings has been explicitly explained. Scatter of data that had happened during the nanoindentation study on MIPS-HAp coatings has been discussed. Further, it has been illustrated how this scatter can be treated in terms of a two-parameter Weibull distribution. MIPS-HAp coatings contain defects, microcracks, and porosity, which finally affect indentation size effect. In this context, the inherent complexity of the indentation size effect has been explained. In addition, the huge importance of fracture toughness evaluation at the microstructural length scale has been highlighted. More stress has been put to indentation-based fracture toughness measurement in general, and nanoindentation-based fracture toughness measurement in particular. Finally, the scenario of microstructure-crack interaction has been depicted to explain why MIPS-HAp coatings showed toughness comparatively higher than those of the HAp coatings deposited by the MAPS process.

In Chapter 8, tribological properties of HAp coatings will be demonstrated because the intended application of such a HAp coating, especially the MIPS-HAp coatings, will involve a dynamic contact situation in the implanted condition. The tribological characteristics, e.g., coefficient of friction, etc., and microstructural evaluations following tribological studies of HAp coatings in general and MIPS-HAp coatings in particular will be discussed next. Results derived from both single-pass scratch tests and reciprocatory wear tests of HAp coatings will be documented in detail in Chapter 8.

References

1. Oliver W. C. and Pharr G. M. 1992. An improved technique for determining hardness and elastic modulus using load and displacement sensing indentation experiments. *Journal of Materials Research*, 7: 1564–1583.
2. Ling Z. and Hou J. 2007. A nanoindentation analysis of the effects of microstructure on elastic properties of Al_2O_3/SiC composites. *Composites Science and Technology*, 67: 3121–3129.
3. Zhu D., Hongna D., Luo F., and Zhou W. 2007. Preparation and mechanical properties of C/C-SiC composites. *Materials Science Forum*, 546–549: 1501–1504.
4. Guicciardi S., Balbo A., Sciti D., Melandri C., and Pezzotti G. 2007. Nanoindentation characterization of SiC-based ceramics. *Journal of the European Ceramic Society*, 27: 1399–1404.
5. Gross K. A., Ray N., and Rokkum M. 2002. The contribution of coating microstructure to degradation and particle release in hydroxyapatite coated prostheses. *Journal of Biomedical Materials Research Part B*, 63, 106–114.
6. Chung S. S., Lee C. K., Hong K. S., and Yoon H. J. 1996. Difference of bonding behavior between four different kinds of hydroxyapatite plate and bone. In *Bioceramics*, Sedel L. and Rey C. (Eds.), vol. 10. UK: Elsevier, 83–86.
7. Kweh S. W. K., Khor K. A., and Cheang P. 2000. Plasma-sprayed hydroxyapatite (HA) coatings with flame-spheroidized feedstock: microstructure and mechanical properties. *Biomaterials*, 21: 1223–1234.
8. Mancini C. E., Berndt C. C., Sun L., and Kucuk A. 2001. Porosity determinations in thermally sprayed hydroxyapatite coatings. *Journal of Materials Science*, 36: 3891–3896.
9. Nieh T. G., Jankowsk A. F., and Koike J. 2001. Processing and characterization of hydroxyapatite coatings on titanium produced by magnetron sputtering. *Journal of Materials Research*, 16: 3238–3245.
10. Nieh T. G., Choi B. W., and Jankowski A. F. 2001. Synthesis and characterization of porous hydroxyapatite and hydroxyapatite coatings. Report submitted to Minerals, Metals and Materials Society Annual Meeting and Exhibition, Los Angeles.
11. Khor K. A., Li H., and Cheang P. 2003. Characterization of the bone-like apatite precipitated on high velocity oxy-fuel (HVOF) sprayed calcium phosphate deposits. *Biomaterials*, 24: 769–775.
12. Cheng G. J., Pirzada D., Cai M., Mohanty P., and Bandyopadhyay A. 2005. Bioceramic coating of hydroxyapatite on titanium substrate with Nd-YAG laser. *Materials Science and Engineering C*, 25: 541–547.
13. Chen Y., Zhang Y. Q., Zhang T. H., Gan C. H., Zheng C. Y., and Yu G. 2006. Carbon nanotube reinforced hydroxyapatite composite coatings produced through laser surface alloying. *Carbon*, 44: 37–45.
14. Arias J. L., Mayor M. B., Pou J., Leng Y., Leon B., and Perez-Amora M. 2003. Micro- and nano-testing of calcium phosphate coatings produced by pulsed laser deposition. *Biomaterials*, 24: 3403–3408.
15. Gross K. A. and Samandari S. S. 2007. Nano-mechanical properties of hydroxyapatite coatings with a focus on the single solidified droplet. *Journal of Australian Ceramic Society*, 43: 98–101.

16. Gu Y. W., Khor K. A., and Cheang P. 2003. In vitro studies of plasma-sprayed hydroxyapatite/Ti-6Al-4V composite coatings in simulated body fluid (SBF). *Biomaterials*, 24: 1603–1611.
17. Mohammadi Z., Moayyed A. A. Z., and Mesgar A. S. M. 2007. Adhesive and cohesive properties by indentation method of plasma-sprayed hydroxyapatite coatings. *Applied Surface Science*, 253: 4960–4965.
18. Khor K. A., Gu Y. W., Pan D., and Cheang P. 2004. Microstructure and mechanical properties of plasma sprayed HA/YSZ/Ti-6Al-4V composite coatings. *Biomaterials*, 25: 4009–4017.
19. Khor K. A., Gu Y. W., Quek C. H., and Cheang P. 2003. Plasma spraying of functionally graded hydroxyapatite/Ti-6Al-4V coatings. *Surface and Coatings Technology*, 168: 195–201.
20. Gross K. A. and Samandari S. S. 2009. Nanoindentation on the surface of thermally sprayed coatings. *Surface and Coatings Technology*, 203: 3516–3520.
21. Dey A., Mukhopadhyay A. K., Gangadharan S., Sinha M. K., Basu D., and Bandyopadhyay N. R. 2009. Nanoindentation study of microplasma sprayed hydroxyapatite coating. *Ceramics International*, 35: 2295–2304.
22. Dey A., Mukhopadhyay A. K., Gangadharan S., Sinha M. K., and Basu D. 2009. Weibull modulus of nano-hardness and elastic modulus of hydroxyapatite coating. *Journal of Materials Science*, 44: 4911–4918.
23. Dey A. and Mukhopadhyay A. K. 2010. Anisotropy in nano-hardness of microplasma sprayed hydroxyapatite coating. *Advances in Applied Ceramics*, 109: 346–354.
24. Wen J., Leng Y., Chen J., and Zhang C. 2000. Chemical gradient in plasma-sprayed HA coatings. *Biomaterials*, 21: 1339–1343.
25. Zhang C., Leng Y., and Chen J. 2001. Elastic and plastic behavior of plasma-sprayed hydroxyapatite coatings on a Ti-6Al-4V substrate. *Biomaterials*, 22: 1357–1363.
26. Li H., Khor K. A., and Cheang P. 2002. Young's modulus and fracture toughness determination of high velocity oxy-fuel-sprayed bioceramic coatings. *Surface and Coatings Technology*, 155: 21–32.
27. Zhang S., Wang Y. S., Zeng X. T., Khor K. A., Weng W., and Sun D. E. 2008. Evaluation of adhesion strength and toughness of fluoridated hydroxyapatite coatings. *Thin Solid Films*, 516: 5162–5167.
28. Xu H. and Pharr G. M. 2006. An improved relation for the effective elastic compliance of a film/substrate system during indentation by a flat cylindrical punch. *Scripta Materialia*, 55: 315–318.
29. Kumar R. R. and Wang M. 2002. Modulus and hardness evaluations of sintered bioceramic powders and functionally graded bioactive composites by nano-indentation technique. *Materials Science and Engineering A*, 338: 230–236.
30. Wallace J. S. and Llavsky J. 1998. Elastic modulus measurements in plasma sprayed deposits. *Journal of Thermal Spray Technology*, 7: 521–526.
31. Guo S. and Kagawa Y. 2004. Young's moduli of zirconia top-coat and thermally grown oxide in a plasma-sprayed thermal barrier coating system. *Scripta Materialia*, 50: 1401–1406.
32. Rice R. W. 1977. Microstructure dependence of mechanical behaviour of ceramics. In *Treatise on materials science and technology II*, MaCrone R. C. (Ed.). New York: Academic Press, 200.

33. Dey A. and Mukhopadhyay A. K. 2014. In-vitro dissolution, microstructural and mechanical characterizations of microplasma sprayed hydroxyapatite coating. *International Journal of Applied Ceramic Technology*, 11: 65–82.
34. Evans A. G., Lawn B. R., and Marshall D. B. 1980. Elastic/plastic indentation damage in ceramics: the median/radial crack system. *Journal of the American Ceramic Society*, 63: 574–581.
35. Venkataraman R. and Krishnamurthy R. 2006. Evaluation of fracture toughness of as plasma sprayed alumina–13 wt.% titania coatings by micro-indentation techniques. *Journal of the European Ceramic Society*, 26: 3075–3081.
36. Zhang S., Sun D., Fu Y., and Du H. 2005. Toughness measurement of thin films: a critical review. *Surface and Coatings Technology*, 198: 74–84.
37. Balani K., Anderson R., Laha T., Andara M., Tercero J., Crumpler E., and Agarwal A. 2007. Plasma-sprayed carbon nanotube reinforced hydroxyapatite coatings and their interaction with human osteoblasts in vitro. *Biomaterials*, 28: 618–624.
38. Li H., Khor K. A., and Cheang P. 2002. Titanium dioxide reinforced hydroxyapatite coatings deposited by high velocity oxy-fuel (HVOF) spray. *Biomaterials*, 23: 85–91.
39. Fu L., Khor K. A., and Lim J. P. 2002. Effects of yttria-stabilized zirconia on plasma-sprayed hydroxyapatite/yttria-stabilized zirconia composite coatings. *Journal of the American Ceramic Society*, 85: 800–806.
40. Fu L. K., Khor A., and Lim J. P. 2001. Processing, microstructure and mechanical properties of yttria stabilized zirconia reinforced hydroxyapatite coatings. *Materials Science and Engineering A*, 316: 46–51.
41. Fu L. K., Khor A., and Lim J. P. 2000. Yttria stabilized zirconia reinforced hydroxyapatite coatings. *Surface and Coatings Technology*, 127: 66–75.
42. Wang M., Yang X. Y., Khor K. A., and Wang Y. 1999. Preparation and characterization of bioactive monolayer and functionally graded coatings. *Journal of Materials Science: Materials in Medicine*, 10: 269–273.
43. Dey A. and Mukhopadhyay A. K. 2011. Fracture toughness of microplasma sprayed hydroxyapatite coating by nanoindentation. *International Journal of Applied Ceramic Technology*, 8: 572–590.
44. Hench L. L. 1998. Bioceramics. *Journal of the American Ceramic Society*, 81: 1705–1733.
45. Kobayashi S., Kawai W., and Wakayama S. 2006. The effect of pressure during sintering on the strength and the fracture toughness of hydroxyapatite ceramics. *Journal of Materials Science: Materials in Medicine*, 17: 1089–1093.
46. Hulbert S. F., Hench L. L., Forbers D., and Bowman L. S. 1982–1983. History of bioceramics. *Ceramurgia International*, 8–9: 131–140.
47. Willmann G., 1999. Coating of implants with hydroxyapatite material connections between bone and metal. *Advanced Engineering Materials*, 1: 95–105.
48. Yang C. Y., Lin R. M., Wang B. C., Lee T. M., Chang E., Hang Y. S., and Chen P. Q. 1997. In vitro and in vivo mechanical evaluations of plasma-sprayed hydroxyapatite coatings on titanium implants: the effect of coating characteristics. *Journal of Biomedical Materials Research*, 37: 335–345.

8

Tribological Properties
of Hydroxyapatite Coatings

8.1 Introduction

In this chapter we discuss the tribological properties of the hydroxyapatite (HAp) coatings. The practical challenges of the biomedical application of a conventional dense high-power MAPS-HAp coating include (1) occlusion of the porous surface, (2) uncontrolled bioresorption, and (3) late delamination with formation of particulate debris [1–4]. The last factor is an issue of significant importance in the case of, e.g., hip implants, because particulate debris of HAp might well accelerate the polyethylene wear-induced granulomatous tissue response with an associated bone lysis [1, 5]. This last factor also constitutes the scope of the discussion on issues related to dynamic contact deformation and damage evolution in a HAp coating. It needs to be appreciated that due to repetitive loading during in vivo implantation, the HAp coating surfaces may often experience the micro/nanoscale wear. Classically speaking, this is what tribology is all about. It needs to be recalled, at the same time, that the discussions presented in earlier chapters show that in spite of having about 11–20% porosity, the MIPS-HAp coatings have bonding strength, hardness, elastic modulus, and fracture toughness higher than those of the MAPS-HAp coatings. Therefore, for in vivo applications, the study of the nanotribological characteristics of the HAp coatings in general and MIPS-HAp coatings in particular emerges as the most important issue. This is what is attempted in this chapter.

8.2 What Does the Literature Say?

It is really appreciable that a wealth of literature [2, 6–23] on the tribological characteristics of HAp or HAp composite coatings exists. A critical survey of the same (e.g., Table 8.1), however, points out that there has not been much

TABLE 8.1

Literature Status on Tribological Properties of HAp or HAp composite Coatings

Coating/Substrate: Processing Route	Scratch/Wear Test	Delamination or Critical Load (P_c) and Coefficient of Friction (COF)	Reference
HAp, HAp + CNT/Ti-6Al-4V: LSA (400 W)	Nanoscratch test (ramping load)	COF = 0.35–0.2 (HAp) COF = 0.05–0.4 (HAp + CNT)	[6]
HAp (FGC)/Ti-6Al-4V: MAO + ED	Scratch adhesion test	P_c = 10 N (HAp + TiO$_2$)	[7]
HAp/Ti-6Al-4V: PLD	Microscratch test (ramping load)	P_{c1} = 1.78–4 N P_{c2} = 3.25–6.7 N P_{c3} = 6.37–7.3 N	[8]
HAp/Ti-6Al-4V: LA	Microscratch test (ramping load)	P_c = 0.3–5.7 N	[9]
HAp/Ti-6Al-4V: LA	Nanoscratch test (ramping load)	P_c = 160–230 mN P_c = 180 mN (after immersion of Hanks' solution)	[10]
HAp, FHAp/Ti-6Al-4V: SG	Nanoscratch test (ramping load)	P_c = 470 mN (FHAp) P_c = 360 mN (HAp)	[11]
HAp/Al, Ti alloy: MAPS (40 kW)	Microscratch test (constant loads)	P_c = 12.3 N (HAp/Ti) P_c = 10.3 N (HAp/Al)	[12]
HAp/Ti: ED	Microscratch test (ramping load)	P_c = 20 N	[13]
HAp/Ti: Biomimetic	Nanoscratch test (ramping load)	Critical pressure = 2.4 GPa Adhesion strength = 34.5 MPa	[14]
HAp, HAp + CNT/Ti-6Al-4V: MAPS	Pin on disc (zirconia pin)	Wear rate: 60.15 g.m^{-2} (HAp) 38.92 g.m^{-2} (HAp + CNT)	[15]
HAp/Ti: WCD	Microscratch test (constant load)	P_c = 13.1 N	[16]
CaP/Ti: MS	Microscratch test (ramping load)	P_c = 15.8–6.4 N	[17]
HAp/Ti-6Al-4V: IBSD, IBAD	Microscratch test (constant load)	P_c = 660 gf (IBSD) P_c = 1050 gf (IBAD)	[18]

Material/system	Test	Results	Ref.
HAp/SS316L: MIPS (1.5 kW), MAPS (40 kW)	Nanoscratch test (constant load) Nanoscratch test (ramping load) Microscratch test (constant loads)	COF = 0.15 (MIPS) COF = 0.5 (MIPS) COF = 0.4–0.6 (as sprayed, MIPS) COF = 0.3–0.4 (polished, MIPS) COF = 0.2–0.4 (as sprayed, MAPS) COF = 0.2 (polished, MAPS) COF-0.5–0.8 (as sprayed, MIPS, after SBF immersion) no delamination both before and after SBF immersion (up to 10.6 N)	[2, 19]
CHAp/Ti: PLD	Microscratch test (ramping load)	$P_c = 0.6$–9.8 N	[20]
HAp/Ti, Si: MS	Nanoscratch test (ramping load)	No delamination	[21]
HAp/Ti-6Al-4V: MAPS (40 kW)	Fretting wear (RT and humidity ~ 80%, frequency ~ 5 Hz (fixed), loads = 5 N, 10 N, oscillatory amplitudes = 50, 100, and 200 μm)	COF = 0.7–0.8 (dry)	[22]
HAp/Ti-6Al-4V: MAPS (40 kW)	Fretting wear (RT and humidity ~ 80%, frequency ~ 5 Hz, loads = 5 N, 10 N, oscillatory amplitudes = 50, 100, and 200 μm)	COF = 0.7–0.8 (dry) COF = 0.2–0.3 (lubrication by bovine albumin)	[23]

Note: CHAp = carbonated HAp, FGC = functionally graded coating, LSA = laser surface alloying, MAO = microarc discharge oxidation, ED = electrophoretic deposition, PLD = pulsed laser deposition, LA = laser ablation technique, FHAp = fluoridated HAp, SG = sol-gel method, MAPS = macroplasma spraying technique, MIPS = microplasma spraying technique, ED = electrochemical deposition, MS = magnetron sputtering method, WCD = wet chemical deposition, CaP = calcium phosphate compound, IBSD = ion beam sputtering deposition, IBAD = ion beam-assisted deposition, nanoscratch test = scratch test in the range of 0–1000 mN load, microscratch test = scratch test in the range of 0–100 N load, SBF = simulated body fluid.

of a systematic study at the micro/nanometric load range; i.e., our knowledge base about the mechanisms of wear processes happening at the micro- or submicrostructural length scale is, to date, far from comprehensive.

Tribological properties of HAp or HAp composite coating were evaluated by the micro- and nanoscratch tests, as well as by the more conventional pin on disc technique (Table 8.1) [2, 6–23]. The coefficients of friction (COFs) of the HAp coatings prepared by the LSA [6], microplasma spraying (MIPS) [2, 19], and macroplasma spraying (MAPS) [2] methods were reported to be in the range of 0.2–0.8. Chen et al. [6] also showed a marginal improvement of the COF after incorporation of a CNT. Similarly, wear properties measured by the pin on disc method were also improved after addition of CNT as the second phase in HAp coatings prepared by the MAPS technique [15]. Further, the fluoridated HAp coatings showed a critical load much higher than that recorded for the pure HAp coating deposited by the sol-gel technique [11].

Nieh et al. [21] and Zhang [12] showed that the degree of adhesion was influenced by the nature of substrate materials. Nieh et al. [21] reported that the HAp-Ti interface is stronger than the HAp-Si interface when the thin films of HAp were deposited by the sputtering method on Ti and Si substrates.

Similarly, a thick HAp coating deposited by the MAPS method on Ti offered a critical load higher than that recorded for a thick HAp coating deposited by the MAPS method on Al [12]. However, the amounts of systematic, in-depth, single-pass scratch-and-wear studies on HAp coatings are far from significant.

It is in this context that both Dey [2] and Dey and Mukhopadhyay [19] reported a significant amount of basic scientific studies about the tribological characteristics of MIPS-HAp coatings. Thus, it was demonstrated [1, 2, 19] that the COF greatly depends on the processing technique (MIPS vs. MAPS), surface roughness (as deposited vs. polished), and load regime under investigation (nano- vs. microscratch test). The COF value of MIPS-HAp coating was 0.15 at 100 µN load. However, COF was increased to 0.5 at 700 mN. Further, an increase in load, e.g., up to 10.6 N, had no significant influence on COF (e.g., ~ 0.6). Moreover, after SBF immersion for 14 days the COF further increased to 0.8, but no delamination was observed [2, 19]. The COFs of MIPS-HAp coatings were always higher than those of the MAPS-HAp coatings, and the COFs of the as-sprayed MIPS- or MAPS-HAp coatings were slightly higher than those of the polished coatings.

Further, Fu et al. [22, 23] conducted fretting wear in both dry and lubrication conditions with bovine albumin at room temperature (RT) and humidity of ~80%. Two loads were employed as 5 and 10 N. The COFs in dry condition (0.7–0.8) were much higher than those realized (e.g., 0.2–0.3) under the lubricated condition, as expected.

Thus, if we critically look into the literature status, we can depict that most of these reports mentioned above (Table 8.1) involve mainly MAPS-HAp coating. The work on MIPS is almost significantly less. This particular need provided the genesis of the present work to emphasize the tribology of MIPS-HAp coatings more.

8.3 Nanoscratch Testing of MIPS-HAp Coatings at Lower Load

The nanoscratch tests were conducted at a speed of 0.14 μm.s^{-1} at room temperature (30°C) with a 90° conical diamond probe that moved over a MIPS-HAp coating surface that was held stationary. The scratch probe had a tip radius of about 1 μm. A commercial machine (Tribo-Indenter, Hysitron, USA) was used for this purpose. The scratch length was 6 μm. The normal load applied on the coating surface was 100 μN. The normal force was constant all along the length of the scratch (Figure 8.1a). The final normal

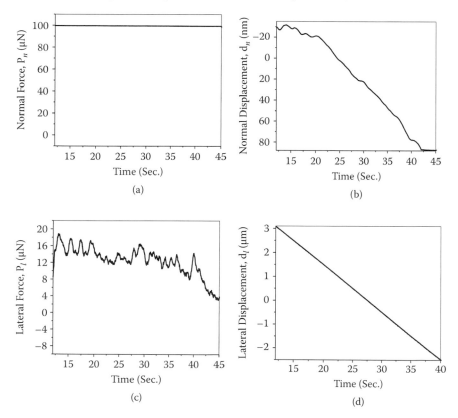

FIGURE 8.1
Data obtained for the MIPS-HAp coating as a function of scratch time during the 100 μN constant normal force nanoscratch experiments: (a) normal force, (b) normal displacement, (c) lateral force, and (d) lateral displacement.

displacement was about 100 nm (Figure 8.1b). The average lateral force had decreased with time (Figure 8.1c). This was probably caused by the adherent presence of the coating material at the interface between the indenter tip and the coating. The trend of the lateral displacement data (Figure 8.1d) followed that of the lateral force data. The conjecture made above is well supported by the small magnitude (~0.15) of the coefficients of friction (Figure 8.2). Finally, topographical (Figure 8.3a) and gradient (Figure 8.3b) in situ scanning probe microscopy (SPM) images of the region on which the

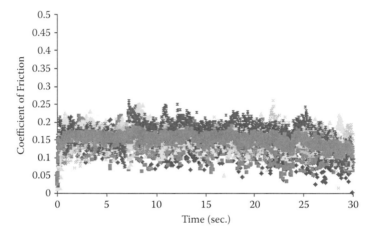

FIGURE 8.2 (See color insert.)
Coefficient of friction of the MIPS-HAp coating as a function of scratch time. Red line represents the average of five scratch test data.

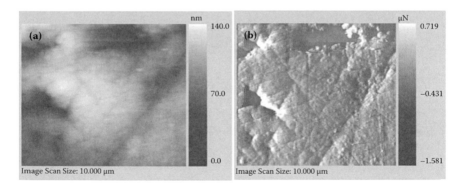

FIGURE 8.3 (See color insert.)
In situ SPM photomicrographs (image scan size: 10×10 μm) of the region of the MIPS-HAp coating on which the nanoscratch experiments were performed (a) in topographical and (b) in gradient mode, showing that no peel-off occurred after the 100 μN constant normal force nanoscratch experiments.

nanoscratch was performed confirmed that no peel-off had occurred. The coating has a microstructure that is characteristically highly heterogeneous in nature. This characteristic heterogeneity is reflected in the typical scatter inherently present in the data.

8.4 Nanoscratch Testing of MIPS-HAp Coating at Higher Load

The same machine mentioned in Section 8.3 was used for this purpose. Here, the normal force was ramped from 0 to 700 mN. These high-load nanoscratch tests were conducted at a speed of 7.69 $\mu m.s^{-1}$ at room temperature (30°C) with a 60° conical diamond probe that moved over a MIPS-HAp coating surface that was held stationary. The scratch probe had a tip radius of about 20 μm. The scratch length was 100 μm. There were multiple scratches made with the diamond tip. It moved in one direction only. Any sudden change in load represents microcracking or delamination of the coating under investigation. The load value at which such a situation happens is taken as the critical load. Conceptually, it signifies the scratch resistance of the MIPS-HAp coating under study. The various nanotribolical data obtained from these experiments are shown in Figure 8.4a–d. The results presented in Figure 8.4a–d suggested that the coating had a critical load of about 400 mN. The coefficient of friction (0.45–0.55) was much higher than that (~0.15) recorded with the low-load scratch test. The microstructure of the scratched coating (Figure 8.5b) was as good as that of the original, unscratched coating (Figure 8.5a). Except for the generation of some microcracks (Figure 8.5b), there was no large-scale peel-off present on the coating surface.

8.5 Microscratch Testing of MIPS-HAp Coatings

These microscratch experiments were also conducted at room temperature (30°C) in a commercial scratch tester (Model TR 102-M3, Ducom, Bangalore, India). The scratch tests were conducted at 2–2.6 N, 5–5.6 N, and 10–10.6 N normal ramping loads with a 90° conical diamond having a tip radius of 200 μm. The scratch length was 3 mm. The scratch speed was 0.2 $mm.s^{-1}$. The scratch offset was 1 mm. In these experiments, however, the sample stage holding the sample moved at the predesignated speed below the indenter statically held in the machine. The scratch tests were conducted on both as-deposited and polished MIPS-HAp coatings.

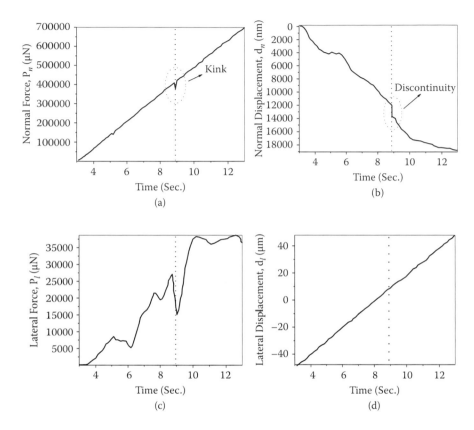

FIGURE 8.4
Data obtained for the MIPS-HAp coating as a function of scratch time during nanoscratch experiments by ramping the normal force from 0 to 700 mN: (a) normal force, (b) normal displacement, (c) lateral force, and (d) lateral displacement. (Reprinted/modified from Dey et al., *Journal of Thermal Spray Technology*, 18: 578–592, 2009. With permission from Springer.)

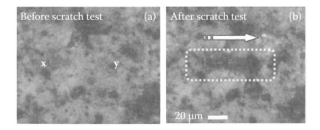

FIGURE 8.5 (See color insert.)
Optical micrographs of the MIPS-HAp coating (a) before (starting and end points of the scratch test marked *x* and *y*, respectively) and (b) after (the arrow indicates the direction of scratching) the nanoscratch testing by ramping the normal force from 0 to 700 mN. (Reprinted/modified from Dey et al., *Journal of Thermal Spray Technology*, 18: 578–592, 2009. With permission from Springer.)

8.5.1 As-Sprayed MIPS-HAp Coating

The data on variations of normal ramping load (P_{nr}) as a function of sliding distance (d) for the as-sprayed MIPS-HAp coating on SS316L are shown in Figure 8.6a. Three different normal ramping loads (e.g., P_{nr} = 2–2.6 N, 5–5.6 N, and 10–10.6 N) were chosen in these experiments. Further, the data on the variations of lateral force (P_l) as a function of sliding distance are shown in Figure 8.6b. All the lateral forces were increased sharply up to a certain sliding distance. These critical values of scratch lengths made by different P_{nr}, of e.g., 2–2.6 N, 5–5.6 N, and 10–10.6 N, were about 0.28, 0.32, and 0.85 mm, respectively. The corresponding P_l values were ~1.5, ~2, and ~5.7 N. These data would suggest the presence of a typical running-in period in terms of sliding distance. Beyond these sliding distances for a given P_{nr}, the value of lateral forces was more or less constant within a band. However, within that band the strong local variations of the data were noticeable. The present MIPS-HAp coating was deposited as a layer-by-layer splat on the metallic substrate, and it had an extremely heterogeneous microstructure consisting of microcracks, pores, unmelted particles, etc. The interaction of the indenter tip with splat boundaries, micropores, microcracks, and other defects possibly caused

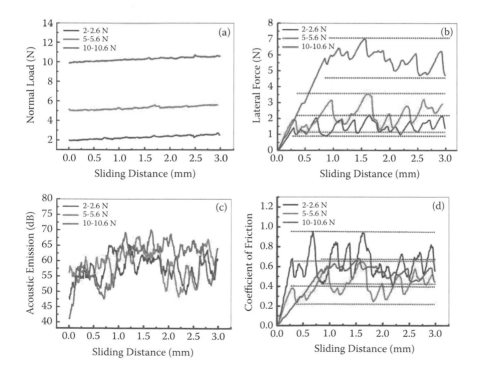

FIGURE 8.6 (See color insert.)
Data on the variations of (a) normal ramping load, (b) lateral force, (c) acoustic emission, and (d) coefficient of friction as a function of sliding distance for the as-sprayed MIPS-HAp coating.

the local variation in the data. The variations of acoustic emissions (AEs) as a function of sliding distance are shown in Figure 8.6c. The AE values of all three cases were found as more or less constant within a band starting from ~48 to 68 dB. Further, the data on variation of COF as a function of sliding distance are shown in Figure 8.6d. Coefficient of friction data at three different ramping loads were increased sharply up to a certain sliding distance. These critical values of scratch lengths made by different P_{nr}, of e.g., 2–2.6 N, 5–5.6 N, and 10–10.6 N, were 0.28, 0.32, and 0.85 mm, respectively. However, such lengths were typically of the order of or less than 10% of the total distance (e.g., 3 mm) slid by the diamond indenter, except at 10 N or higher loads, where it was about 30% of the distance slid. These data would also suggest the presence of a running-in period in terms of the sliding distance. The corresponding COF values were ~0.7, ~0.4, and ~0.56. Afterwards, the value of COF was almost constant within a band. However, within that regime strong local variations of the data were noticeable. This could be explained in terms of variation of the corresponding lateral force data.

Optical photomicrographs of the scratches made at different ramping normal loads, of e.g., P_{nr} = 2–2.6 N, 5–5.6 N, and 10–10.6 N, are shown in Figure 8.7a–c, respectively. The impression of scratch was more pronounced

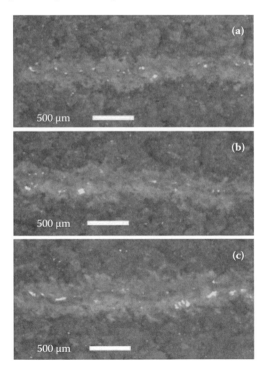

FIGURE 8.7
Optical photomicrographs of scratches in the as-sprayed MIPS-HAp coating at three different normal ramping loads: (a) 2–2.6 N, (b) 5–5.6 N, and (c) 10–10.6 N.

as the normal load was increased. In addition, no delaminations or macro-fractures were observed for the aforesaid three cases.

Further, scanning electron microscopy (SEM) photomicrographs of the scratches at ramping normal load, e.g., $P_{nr} = 10\text{--}10.6$ N, are shown in Figure 8.8. No delaminations or macrofractures were observed up to $P_{nr} = 10\text{--}10.6$ N. However, microcracks followed by microfracture were produced in the path of the scratch track. Further, the edges of the scratch grooves were wavy in all the cases examined. This could be due to the large porosity levels in the MIPS-HAp coatings. In the case of bulk ceramics, a similar opinion was put forward by Subhash and Bandyo [24]. A higher-magnification SEM photo-micrograph of the scratch track of an as-sprayed MIPS-HAp at 10–10.6 N is shown in Figure 8.9. A bunch of parallel microcracks were formed nearly perpendicular to the scratch direction. These types of microcracks are often observed during scratch testing on the plasma sprayed ceramic coatings as well as bulk ceramics [24–28]. These cracks most likely formed as a result of

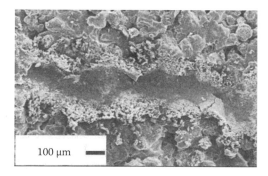

FIGURE 8.8
SEM photomicrographs of scratches in the as-sprayed MIPS-HAp coating at 10–10.6 N.

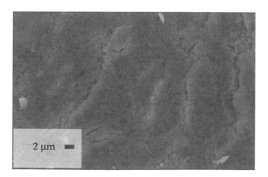

FIGURE 8.9
Higher-magnification SEM photomicrograph of scratch track in the as-sprayed MIPS-HAp coating at 10–10.6 N showing formation of almost parallel microcracks nearly perpendicular to the scratch direction (direction of the scratches were from the left to right).

the tensile frictional stress, which had acted behind the trailing edge of the indenter [26, 28].

8.5.2 Polished MIPS-HAp Coating

The data on the variations of normal ramping load (P_{nr}) as a function of sliding distance (d) for the polished MIPS-HAp coating on SS316L are shown in Figure 8.10a. Three different normal ramping loads (e.g., P_{nr} = 2–2.6 N, 5–5.6 N, 10–10.6 N), as were chosen for the as-sprayed coating, were also chosen in the present experiment. A sharp kink was recorded at a higher ramping load of 10–10.6 N, which possibly suggested occurrence of a microchipping or delamination at a local microstructural length scale during the microscratch testing. However, the coating was not totally delaminated, even with further loading after the microchipping. Further, the data on the variations of lateral force (P_l) as a function of sliding distance are shown

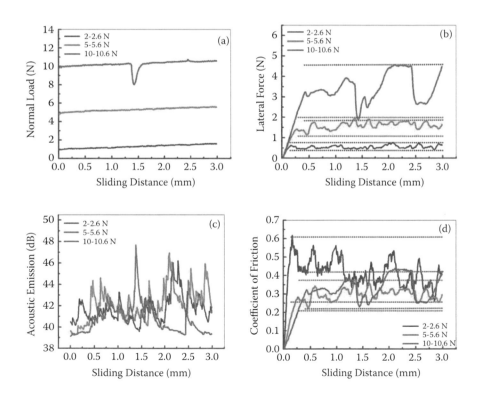

FIGURE 8.10 (See color insert.)
Data on the variations of (a) normal ramping load, (b) lateral force, (c) acoustic emission, and (d) coefficient of friction as a function of sliding distance for the polished MIPS-HAp coating.

in Figure 8.10b. All the lateral forces increased sharply up to a certain slid-ing distance. These critical values of scratch lengths made by different P_{nr}, e.g., 2–2.6 N, 5–5.6 N, and 10–10.6 N, were 0.16, 0.28, and 0.44 mm, respec-tively. The corresponding P_l values were ~0.62, ~1.4, and ~3.2 N. These data would suggest the presence of a typical running-in period in terms of the sliding distance. Beyond these sliding distances for a given P_{nr}, the value of lateral forces was more or less constant within a band. However, within that band, strong local variations of the data were noticeable. The interaction of the indenter tip with splat boundaries, micropores, microcracks, and other defects (i.e., characteristic features of plasma sprayed coating, as discussed in Chapter X) possibly caused the local variation of data. These values were comparatively much lower than those recorded for the as-sprayed MIPS-HAp coating discussed earlier. However, the variations of lateral force data at 10–10.6 N normal load were much higher due to local delamination and microchipping (Figure 8.10b). The variations of AEs as a function of sliding distance are shown in Figure 8.10c. The AE values of all three cases were found to remain more or less constant within a band starting from ~39 to 48 dB. These values of AE were much lower than those of the as-sprayed coating presumably, due to better surface finish or a comparatively lower R_a value. Further, the data on variations of the COF as a function of sliding distance are shown in Figure 8.10d. Coefficients of friction data at three dif-ferent ramping loads were increased sharply up to a certain sliding distance. These critical values of scratch lengths made by different P_{nr}, e.g., 2–2.6 N, 5–5.6 N, and 10–10.6 N, were 0.16, 0.27, and 0.44 mm, respectively. However, such lengths were typically about 5, 10, and 15% of the total distance (e.g., 3 mm) slid by the diamond indenter. These data would also suggest the pres-ence of a running-in period in terms of the sliding distance. The correspond-ing COF values were ~0.62, ~0.27, and ~0.31. Afterward, the value of COF was almost constant within a band. However, within that band, the strong local variations of the data were quiet noticeable. This could be explained in terms of the variation of the corresponding lateral force data.

Optical photomicrographs of the scratches made at different ramp-ing normal loads, e.g., P_{nr} = 2–2.6 N, 5–5.6 N, and 10–10.6 N, are shown in Figure 8.11a–c, respectively. The impression of scratch was more pronounced as the normal load was increased. In addition, no major delaminations or macrofractures were observed for the aforesaid three cases. However, the signature of a local delamination or microchipping was recorded in case of higher load scratching, e.g., 10–10.6 N ramping normal load (Figure 8.11c).

Further, a SEM photomicrograph of the scratch at ramping normal load, e.g., P_{nr} = 5–5.6 N, is shown in Figure 8.12. No major macrofractures or delaminations were observed for the aforesaid three cases, although some localized delamination or microchipping of the MIPS-HAp occurred at the highest ramping normal load of 10–10.6 N used in the present experiments.

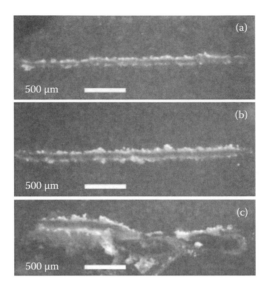

FIGURE 8.11
Optical photomicrographs of scratches on the polished MIPS-HAp coating at three different ramping normal loads: (a) 2–2.6 N, (b) 5–5.6 N, and (c) 10–10.6 N.

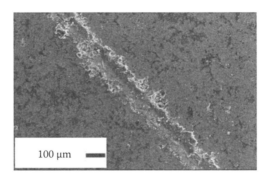

FIGURE 8.12
SEM photomicrographs of scratch on the polished MIPS-HAp coating at 5–5.6 N.

8.6 Microscratch Testing of MIPS-HAp Coatings before and after the SBF Immersion

The data on variations of the COF as a function of sliding distance both before and after immersion in the SBF solution are shown in Figure 8.13. All the experimentally measured data on the COF had increased sharply up to a certain sliding distance. These data would also suggest the presence of a

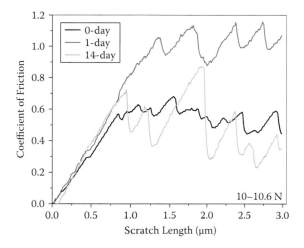

FIGURE 8.13 (See color insert.)
Variations of the coefficient of friction (COF) of the MIPS-HAp coating as a function of slid-ing distance before and after immersion in the SBF solution. (Reprinted/modified from Dey and Mukhopadhyay, *International Journal of Applied Ceramic Technology*, 11: 65–82, 2013. With permission from Wiley.)

running-in period in terms of the sliding distance. Afterward, the value of COF remained almost constant within a band. However, within that band the strong local variations of data were noticeable. Moreover, the variation in the case of the samples immersed in the SBF solution was much higher than that of the virgin MIPS-HAp coating samples. The interaction of the moving indenter tip with deposited loose apatite layer after immersion in the SBF solution possibly caused the local variations of the data.

The value of COF was initially increased after just 1 day of immersion in SBF solution. This happened possibly because much more dissolution took place instead of apatite deposition at the early time of immersion in the SBF solu-tion, as discussed in Chapter 5. Afterward, the average value of COF was dropped to 0.46 after 14 days of immersion in the SBF solution. Surprisingly, this value was close to the value of COF measured for the virgin coating. This happened possibly because the deposition of loose apatite increased as the immersion time progressed from 1 day to 14 days, as discussed in Chapter 5. Further, those loose apatites could produce microwear debris when they made contact with the indenter tip. Therefore, the possibility of the contacts of the indenter with the surface asperities could be much higher [29]. If such wear debris is engulfed between the indenter sides and the coating surface, the average coefficient of friction should drop, as was indeed experimentally observed. In other words, after 14 days of immersion in the SBF solution, the MIPS-HAp samples showed an excellent surface mechanical property, e.g., better scratch resistance, even when scratched at a ramping load of 10–10.6 N.

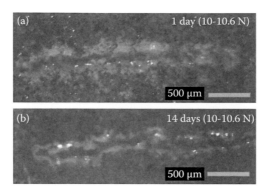

FIGURE 8.14
Optical microscopy images of the scratch paths of the MIPS-HAp coating after the immersion in SBF solution at 10–10.6 N ramping load: (a) 1 day and (b) 14 days.

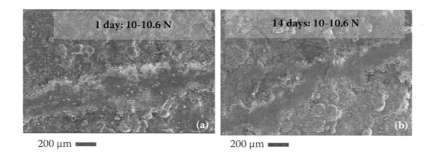

FIGURE 8.15
SEM images of the scratch paths of the MIPS-HAp coating after the immersion in SBF solution at 10–10.6 N ramping load: (a) 1 day and (b) 14 days.

The detailed microstructural features of the scratch track are shown in optical micrographs (Figure 8.14a, b) and scanning electron microscopy photomicrographs (Figure 8.15a, b). No delamination or coating peeling off had been found. Therefore, these data provided conclusive proof regarding the reliability of the coating adhesion during in vitro test.

8.7 Summary

The tribological properties investigated by both scratch and fretting wear tests of HAp coatings were described. Nanoscratch tests on the MIPS-HAp coating with both constant and ramping loads showed no signature of macrofracture, large-scale delamination, or peel-off, thus possibly corroborating the presence of a strain-tolerant microstructure. The microscratch test on the MIPS-HAp coating at ramping normal loads usually showed no large-scale delamination or coating peel-off except at very high loads, e.g., 10–10.6 N. The average COF was about ~0.4 for as-sprayed, but reduced further for the polished MIPS-HAp coating. In general, COF of the as-sprayed MIPS-HAp coating was slightly higher than that of the polished MIPS-HAp coating, due to higher surface roughness (R_a ~ 0.4 µm) and porosity (p ~ 20%) of the former. The values of the coefficient of friction for the MIPS-HAp coating were SBF immersion time dependent. However, the microscratch test at a relatively higher ramping load (e.g., 10–10.6 N) showed no large-scale delamination or coating peel-off, which proved the stability of the coating after immersion in synthetically produced body fluid environment.

Introduction of residual stress in HAp coatings during deposition time will be discussed next in Chapter 9. The origin of it, common methodologies to evaluate residual stress, and their relative merits and demerits will be described in detail. The residual stress of HAp coatings developed by both high- and low-temperature deposition techniques, like plasma sprayed and sol-gel-based HAp coatings, will be also discussed in the next chapter. Further, the utilization of a novel nanoindentation technique for measuring local residual stress of MIPS-HAp coatings will be highlighted.

References

1. Dey A., Mukhopadhyay A. K., Gangadharan S., Sinha M. K., and Basu D. 2009. Characterization of microplasma sprayed hydroxyapatite coating. *Journal of Thermal Spray Technology*, 18: 578–592.
2. Dey A. 2011. Physico-chemical and mechanical characterization of bioactive ceramic coating. PhD dissertation, Indian Institute and Engineering Technology (formerly Bengal Engineering and Science University), Shibpur, Howrah, India.
3. Bauer T. W., Geesink R. C., Zimmerman R., and McMahon J. T. 1991. Hydroxyapatite-coated femoral stems. Histological analysis of components retrieved at autopsy. *Journal of Bone and Joint Surgery*, 73: 1439–1452.
4. Collier J. P., Surprenant V. A., Mayor M. B., Wrona M., Jensen R. E., and Surprenant H. P. 1993. Loss of hydroxyapatite coating on retrieved total hip components. *Journal of Arthroplasty*, 8: 389–393.
5. Rothman R. H., Hozack W. J., Ranawat A., and Moriarty L. 1996. Hydroxyapatite-coated femoral stems. A matched-pair analysis of coated and uncoated implants. *Journal of Bone and Joint Surgery*, 78: 319–324.
6. Chen Y., Zhang T. H., Gan C. H., and Yu G. 2007. Wear studies on hydroxiapatite composite coating reinforced by carbon nanotube. *Carbon*, 45: 998–1004.
7. Nie X., Leyland A., and Matthews A. 2000. Deposition of layered bioceramic HA/TiO₂ coating on Ti alloy using a hybrid technique of micro-arc deposition and electrophoresis. *Surface and Coatings Technology*, 125: 407–414.

8. Blind O., Klein L. H., Dailey B., and Jordan L. 2005. Characterization of hydroxy-apatite films obtained by pulsed-laser deposition on Ti and Ti-6AL-4V substrates. *Dental Materials*, 21: 1017–1024.

9. Fernandez-Pradas J. M., Cleries L., Martinez E., Sardin G., Esteve J., and Morenza J. L. 2001. Influence of thickness on the properties of HA coating deposited by KrF laser ablation. *Biomaterials*, 22: 2171–2175.

10. Cleries L., Martinez E., Fernandez-Pradas J. M., Sardine G., Esteve J., and Morenza J. L. 2000. Mechanical properties of calcium phosphate coating deposited by laser ablation. *Biomaterials*, 21: 967–971.

11. Zhang S., Xianting Z., Yongsheng W., Kui C., and Wenjian W. 2006. Adhesion strength of sol-gel derived fluoridated hydroxyapatite coatings. *Surface and Coatings Technology*, 200: 6350–6354.

12. Kozerski S., Pawlowski L., Jaworski R., Roudet F., and Petit F. 2010. Two zones microstructure of suspension plasma sprayed HA coating. *Surface and Coatings Technology*, 204: 1380–1387.

13. Kuo M.C. and Yen S.K. 2002. The process of electrochemical deposited hydroxy-apatite coatings on biomedical titanium at room temperature. *Materials Science and Engineering C*, 20: 153–160.

14. Forsgren J., Svahn F., Jarmar T., and Engqvist H. 2007. Formation and adhesion of biomimetic hydroxyapatite deposited on titanium substrates. *Acta Biomaterialia*, 3: 980–984.

15. Balani K., Chen Y., Hamirkar S. P., Dahotre N. B., and Agarwal A. 2007. Tribological behaviour of plasma sprayed carbon nanotube reinforced HA coating in physiological solution. *Acta Biomaterialia*, 3: 944–951.

16. Rohanizadeh R., Legeros R. Z., Harsono M., and Bendavid A. 2005. Adherent apatite coating on titanium substrate using chemical deposition. *Journal of Biomedical Materials Research Part A*, 72: 428–438.

17. Wolke J. G. C., de Groot K., and Jansen J. A. 1998. Dissolution and adhesion behaviour of radio frequency magnetron sputtered Ca-P coatings. *Journal of Materials Science*, 33: 3371–3376.

18. Cui F. Z., Luo Z. S., and Fleng Q. L. 1997. Highly adhesive hydroxyapatite coatings on titanium alloy formed by ion beam assisted deposition. *Journal of Material Science: Material in Medicine*, 8: 403–405.

19. Dey A. and Mukhopadhyay A. K. 2013. In-vitro dissolution, microstructural and mechanical characterizations of microplasma sprayed hydroxyapatite coating. *International Journal of Applied Ceramic Technology*, 11: 65–82.

20. Arias J. L., Mayor M. B., Pou J., Leng Y., Leon B., and Perez-Amora M. 2003. Micro- and nano-testing of calcium phosphate coatings produced by pulsed laser deposition. *Biomaterials*, 24: 3403–3408.

21. Nieh T. G., Jankowsk A. F., and Koike J. 2001. Processing and characterization of hydroxyapatite coatings on titanium produced by magnetron sputtering. *Journal of Materials Research*, 16: 3238–3245.

22. Fu Y., Batchelor A. W., Wang Y., and Khor K. A. 1998. Fretting wear behaviors of thermal sprayed hydroxyapatite coating under unlubricated conditions. *Wear*, 217: 132–139.

23. Fu Y., Batchelor A. W., Wang Y., and Khor K. A. 1999. Fretting wear behaviors of thermal sprayed hydroxyapatite coating lubricated with bovine albumin. *Wear*, 230: 98–102.

24. Subhash G. and Bandyo R. 2005. A new scratch resistance measure for structural ceramics. *Journal of the American Ceramic Society*, 88: 918–925.
25. Zhang J., Chen H., Lee S. W., and Ding C. X. 2007. Evaluation of adhesion/ cohesion of plasma sprayed ceramic coatings by scratch testing. In *Proceedings of Thermal spray 2007: global coating solutions*, Marple B. R., Hyland M. M., Lau Y. C., Li C. J., Lima R. S., and Montavon G. (Eds.). Materials Park, OH: ASM International, 472–477.
26. Xie Y. and Hawthorne H. M. 1999. Wear mechanism of plasma-sprayed alumina coating in sliding contacts with harder asperities. *Wear*, 225–229: 90–103.
27. Veloso G., Alves H. R., and Branco J. R. T. 2004. Effects of isothermal treatment on microstructure and scratch test behavior of plasma sprayed zirconia coatings. *Materials Research*, 7: 195–202.
28. Arata Y., Ohmori A., and Li C. J. 1998. Fracture behavior of plasma sprayed ceramic coatings in scratch test. *Transactions of JWRI*, 17: 31–35.
29. Hench L. L., Splinter R. J., Allen W. C., and Greenlee T. K. 1971. Bonding mechanisms at the interface of ceramic prosthetic materials. *Journal of Biomedical Materials Research*, 5: 117–141.

9

Residual Stress of Hydroxyapatite Coating

9.1 Introduction

The main idea behind this chapter is to provide a comprehensive picture of the locked-in or residual stresses that exist in the hydroxyapatite (HAp) coating, and then how they can influence the properties of the coating in general, and mechanical properties in particular. Now the first question that we need to ask ourselves is: What is meant by the term *residual stress*?

Well, residual stresses or locked-in stresses are conceptualized as the stresses that remain confined within a given solid even in the absence of any external mechanical or thermal loading conditions. These stresses are generated during manufacturing or processing or during surface finishing treatments.

Depending on the application and its relative magnitudes, such residual stresses may be either beneficial or detrimental for a given solid or coating. For instance, let us consider the case of the toughened glasses. Here, through optimized processing and post-processing steps, a compressive residual stress of large magnitude is judiciously introduced for the purpose of toughening into the tensile surface of a given glass. The presence of this compressive residual stress at the surface must be overcome by the applied stress to make the surface flaws activated to cause failure of the toughened glass. As a result, the surface flaws face much lower tensile stress effectively, and thereby resist catastrophic failure to a large extent. Consequently, the toughness is enhanced. Thus, in this case, the presence of the residual compressive surface stress can and does act as a boon.

On the other hand, we can imagine situations, e.g., wherein a coating or a thin film has a large magnitude of tensile residual stress active at the surface. In the presence of such a large tensile residual stress, only a nominally small magnitude of externally applied tensile stress, then, will make the total magnitude of effective tensile stress so high at the surface that it will easily cause delamination at the interface between the coating and thin film. In such cases, therefore, the presence of residual stress can and does turn out to be detrimental for the structural integrity of a given coating or thin film.

Preparation of plasma sprayed HAp coating on metallic substrate is a high-temperature process. For such processing there are the two main contributors to the overall residual stress: (1) quenching-induced stress and (2) thermal mismatch-induced stress [1]. Now, in-service reliability of the HAp-coated metallic implants in in vivo service is determined by their long-term performance. The same can be degraded due to the presence of residual stress inside the coated implants. Ideally, we would like the residual stress to be nonexistent. But in reality, this is very difficult, if not impossible, to make it happen. No matter what is the amount of effort that we may dedicate, residual stress will always be there in a plasma sprayed coating. As mentioned above, the danger is that if the residual stress has an unmanageably high magnitude, it may often lead to delamination at the interface between the coating and the substrate, and eventually to a premature coating failure. If and when this happens, revision surgery is the only way. This definitely is the most unwanted situation from the patient's point of view, and hence all efforts should be directed to avoid such a situation at any cost. Clearly then, the exploration of the residual stress aspect in HAp coatings in general, and in those processed by plasma spraying in particular, turns out to be an issue of significant importance that merits dedicated discussion. This is what we have presented in the following sections of this chapter.

9.2 Origin of Residual Stress

The plasma spray process involves introduction of the material, in the form of fine powder, to a plasma flame, which melts the particles and propels them toward the substrate to be coated. Upon impact, the particles flatten, cool down, and solidify, forming a solid layer. As a result of this process, the coatings have properties quite different from bulk materials of the same composition, as a consequence of porosity, anisotropy, and residual stress [1–22].

The large temperature differences experienced during the deposition process are responsible for the development of residual stress in such coatings. When the molten particles, i.e., splats, strike the substrate, they are rapidly quenched, while their contraction is constrained by their adherence to the substrate. This leads to tensile/compressive stress in the coating, commonly referred to as the quenching stress [1]. Thermal mismatch stress develops due to the difference in the coefficients of thermal expansions between the coating and the metallic implants. Thus, the quenching and the thermally induced mismatch stresses are the two main contributions to the overall residual stress. High residual stresses often lead to cracking or buckling of the coating, and therefore it is essential to study the nature and extent of the residual stress condition of the coating to ensure its trouble-free long usage as an implant.

It has been well known that residual stress is inherently induced in the coatings prepared by the plasma spraying method. This residual stress is caused by the difference in thermal properties between the coating and the substrate materials, combined with the complicated solidification processes of the coatings.

As mentioned above, the critical evaluation of the residual stress states in the plasma sprayed HAp coatings on metallic substrates for orthopedic implants is very important for the durability of the systems and their long-term successes in applications. The magnitude of the residual stress along the spraying direction is larger than that perpendicular to the spraying direction. This can be rationalized on the basis that the structure parallel to the spraying direction is continuously deposited, but that perpendicular to the spraying direction is discontinuously deposited. It is conjectured, generally, that a continuous structure of coating would result in unrelaxed strain, and hence higher residual stress.

When the coating rapidly solidified after plasma spraying due to the mismatch of thermal expansion coefficients of coating and substrate [16, 17], the residual stresses generally occurred near the interface of the plasma sprayed ceramic coating and the metallic substrate [5, 18, 19].

The thermal expansion coefficient of the metallic substrate is generally higher than that of the ceramic coating; hence, a compressive residual stress is expected to develop after cooling down. Depending on the relative magnitude, the compressive stress may accelerate or decelerate the tendencies of the coatings to debond [19–21] out from the substrate-coating interfaces.

9.3 Identification of Residual Stress and Importance

In vivo tests have demonstrated that under the shear loading condition, the implants could fail in the bone near the HAp coating–bone interface, at the HAp coating–bone interface, between lamellar splats in the plasma sprayed HAp coating, and at the HAp coating–substrate interface [23]. One of the major factors hypothesized to cause the failure of the implants mechanically and physiologically in the body fluid environment was the residual stress, but no evidence was provided [23]. The reason for the importance of residual stress was that the lamellar splat interfaces and HAp coating–substrate interface might be subjected to an induced tensile stress in the through-thickness direction of the coating [13]. This tensile stress was argued to be strong enough to tear the coating apart from the substrate.

9.4 Factors Affecting Residual Stress

Residual stress in the coating might vary with coating thickness [20, 22], spraying parameter, and the temperature of the substrate [16, 17]. In general, residual stress increases with the thickness of the coating and the temperature of the specimen during plasma spraying.

9.5 Common Methodologies to Evaluate Residual Stress

There are several commonly used methods for residual stress determination in coatings. These are:

1. Diffraction methods (e.g., X-ray or neutron)
2. Raman piezospectroscopy
3. Mathematical modeling (e.g., analytical or numerical)
4. Materials removal techniques (e.g., hole drilling, layer removal)
5. Curvature methods (e.g., in situ curvature, deflection, or strain measurement)
6. Nanoindentation

Each technique has certain advantages as well as some inherent limitations [24, 25].

9.6 Relative Advantages and Disadvantages

With regard to the diffraction method, the most popular method for direct measurement of residual stresses is to monitor the shift of selected X-ray diffraction (XRD) peaks [26]. Due to the limited penetration depth of X-ray, an alternative to XRD is the use of neutrons. However, scattering intensities of neutron diffraction tend to be relatively low, so it is very difficult to obtain sufficient data in a reasonable time from small volumes, such as those of interest in coatings.

The diffraction method also includes Raman spectroscopy, which has been used to measure stresses in CVD films. However, this technique has not yet been widely applied to the cases of the thermally sprayed coatings, e.g., MAPS/MIPS-HAp coatings.

The curvature method of the residual stress evaluation continuously measures the curvature of the deposit-substrate couple during the spraying process [27, 28]. Another technique is measuring the apparent curvature and deflection of the intact specimen that had cooled down after the spraying process [29, 30]. The curvature method has the important advantage of being nondestructive and applicable while deposition is taking place.

The hole drilling and nanoindentation are semidestructive test processes in nature, as the holes or indentations are very tiny and may not influence the properties of the specimen. While analytical models are not a direct measurement technique, they can still be used to predict the amount and nature of residual stress present in the specimen.

It is true that each of the aforesaid techniques has its own merits and demerits. Therefore, the appropriate technique must be very judiciously chosen for measuring residual stress. For instance, the XRD-$\sin^2\psi$ technique can be useful to evaluate the residual stress only in a thin layer, assuming that it behaves like a bulk, isotropic material. However, in practical applications, in particular for plasma sprayed HAp coatings, these assumptions do not hold well, as the coating is always heterogeneous in nature. In contrast, when we talk about comparatively thick layers of either a graded or an inhomogeneous composition, the materials removal methods in general and the hole drilling method in particular can be the appropriate method of preference to measure residual stress. However, the hole drilling method remains true for local stress measurement rather than global stress, which can be measured by XRD-based techniques. The spatial resolutions of materials removal methods are much inferior to that which can be attained by the XRD-$\sin^2\psi$-based technique. Furthermore, materials removal methods are capable of simple specimen geometries, while $\sin^2\psi$-based measurement methodology can be applied to complicated geometries. However, in spite of so many advantageous points, the $\sin^2\psi$-based measurement methodology is comparatively more tedious and complicated than the other methods.

In this context, the nanoindentation technique [31–36] has emerged recently as one of the potential methods for evaluation of residual stresses for both bulk and ceramic coatings/thin films. This technique has several merits, and hence is gaining popularity. The foremost advantage of this method is that it is less tedious and complex. Further, it takes only little time to complete a given set of measurements. Another advantage of this method is that it can be employed for the evaluation of either tensile or compressive residual stress, as well as plastic strain.

It is definitely interesting to note, in addition, that the same nanoindentation technique can be and is indeed utilized to estimate both local and volume-averaged global mechanical properties, such as Young's modulus, nanohardness, yield strength, strain-hardening exponent, and tensile strength [36]. Now, it must be realized that residual stress may be active over both short and long ranges. The unique advantage of the nanoindentation technique for measuring residual stress is that the one single general methodology can be exploited to probe both short-range and long-range residual stresses [31, 36].

Therefore, it is not at all surprising to note that both the ease and convenience, as well as the accuracy, of the nanoindentation method make it definitely far superior to the more conventional methods, like hole drilling, layer removal, strain evaluation, displacement/curvature measurements, piezospectroscopy, XRD-based mreasurements, neutron diffraction, etc., as already mentioned above. However, it is imperative to give a brief description of the techniques mentioned above before we can go into the details of the nanoindentation technique.

9.6.1 XRD Technique

The residual stress measurement by XRD technique is usually performed in two ways employing, e.g., (1) Hooke's law and (2) the $\sin^2\psi$ method. In the first method, residual stress, σ_r, should be measured by multiplying strain, ε, and Young's modulus, E. The strain should be measured from the XRD data, comparing between coating data and corresponding powder XRD data. Finally, the residual stress of the HAp coating can be estimated from Equations (9.1) to (9.6), proposed by Brown and his coworkers [5]. The present authors have also estimated the residual stress of the microplasma spraying (MIPS) HAp coating [36] following [5]. In other words, the residual stress can be represented as

$$\sigma_r = E_0 \varepsilon \qquad (9.1)$$

where σ_r is the residual stress of the coating, E_0 is the Young's modulus of the stress-free virgin material, e.g., freestanding coating, and ε is the residual strain, calculated from XRD data by the following relationship:

$$\varepsilon = \frac{(d_c - d_0)}{d_0} \qquad (9.2)$$

where ε was calculated by the ratio of the difference between the interplanner spacing (d_c) of the coating and the interplanner spacing (d_0) of the powder and d_0.

The shift of a peak corresponds to a change in the unit cell dimensions, and this is usually caused by stress. Using the software attached with the XRD machine, the position of a peak is accurately measured. It is important to select an appropriate peak, with a reasonable relative intensity with respect to that of the I_{100} hydroxyapatite peak and having simple *hkl* indices, such as the (300) and (004) planes, which give the stress in the two major planes of the hydroxyapatite structure. Since the hydroxyapatite structure is hexagonal, i.e., $a = b \neq c$, the lattice constants a and c can be calculated from the following relationship [5]:

$$d = \frac{1}{\sqrt{\frac{4}{3a^2}(h^2 + k^2 + hk) + \frac{l^2}{c^2}}} \tag{9.3}$$

For the (300) plane this simplifies to

$$a = 2\sqrt{3}d \tag{9.4}$$

and

$$c = 4d \tag{9.5}$$

for the (004) plane, where d is the interplanner spacing calculated from the XRD pattern using Bragg's law, which rearranges to

$$d = \frac{\lambda}{\sin \theta} \tag{9.6}$$

where λ is the wavelength of the incident beam and θ is the measured diffraction angle. Once the lattice parameters have been determined for the powders and the coatings, the lattice strain can be worked out, and by using the measured modulus for hydroxyapatite, the residual stress in the coating can be calculated.

Brown et al. [5] calculated tensile residual stresses of 200–450 MPa for HAp coatings using Hooke's law. The X-ray diffraction method was used to measure residual stress in HAp coatings made by the air plasma spraying (APS) and high-velocity oxy fuel (HVOF) techniques. Three types of spraying parameters were used: gas mixtures of argon (Ar)/nitrogen (N_2) and argon/helium (He) were used for the APS process, while a propane/oxygen mixture was utilized for the HVOF process. For the coatings obtained with the Ar/N_2 and Ar/He gas mixtures and the HVOF process, the residual tensile stress magnitudes were estimated to be, e.g., ~350–450 MPa, 200–300 MPa, and 70–120 MPa, respectively. Further, it was found that the tensile residual stresses increased in proportion to the coating thickness.

On the other hand, the XRD-based $\sin^2\psi$ method is much more complicated than the first one. Yang and his coworker [8] extensively studied residual stress of plasma sprayed HAp coating by this method. According to them [8], the principal stresses σ_x and σ_y are parallel to the plasma sprayed HAp surface, while σ_z was considered zero, shown schematically in Figure 9.1. The residual stress, $\sigma_{,,}$ should be measured at an unknown angle ϕ with principal stress σ_x. The angle ϕ can be varied from 0° to 90°. The angle ϕ signifies the plasma spraying direction. Thus, $\phi = 0°$ corresponds to a direction that

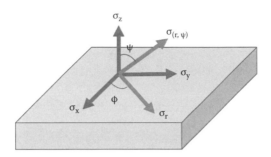

FIGURE 9.1 (See color insert.)
Schematic representation of stresses acting on a residually stressed body. (Adopted/modified and redrawn from Yang et al., *Biomaterials*, 21: 1327–1337, 2000. With permission from Elsevier.)

is parallel to the plasma spraying direction, while $\phi = 90°$ means a direction that is perpendicular to the plasma spraying direction.

Now, it is worth mentioning in this connection that the angle Ψ is known as the angle of tilt to the surface normal of the specimen. Different Ψ angles can be taken. In particular, in the work reported by Yang et al. [8], six Ψ values have been taken as, e.g., 0°, 18.43°, 26.56°, 30°, 33.21°, and 36.27°. Now, the residual stress, σ_r, in the HAp coating acting in the ϕ direction with respect to σ_x can be calculated by Equation (9.7):

$$\sigma_r = \frac{E_0}{(1+v)} \frac{\Delta d/d}{\sin^2 \psi} \tag{9.7}$$

where v is Poisson's ratio of the coating, θ is the diffraction angle, Ψ denotes the tilt angle of the specimen, d is the interplanar spacing (lattice spacing) measured at $\Psi = 0°$ ($d_{\psi=0°}$), and Δd is the difference of $d_{\psi=\psi}$ and $d_{\psi=0°}$ (where $d_{\psi=\psi}$ is the interplanar spacing at $\Psi = \Psi$). The term

$$\frac{\Delta d/d}{\sin^2 \psi} ,$$

usually known as m, is determined from a least-squares fit of the slope derived from the plot of $\Delta d/d$ against $\sin^2\psi$.

Strictly speaking, E_0 should be taken from X-ray elastic constants rather than from nanoindentation measurements. However, in the absence of any reported X-ray elastic constant data, the Young's modulus data experimentally measured by the nanoindentation or three/four-point bending test or any other standard method, e.g., use of ultrasonic pulse echo technique, flexural resonance, etc., can be used. It is, of course, worth mentioning that other researchers [3, 4, 8, 12] had used the data of elastic modulus measured by the three-point bending test on a freestanding coating for residual stress measurement of the HAp coating by the X-ray based $\sin^2\psi$ method. Nevertheless, the present authors employed the nanoindentation technique to measure Young's modulus data of the freestanding HAp coating [36].

Millet et al. [11] obtained compressive residual stresses of ~12.5–29 MPa in HAp coatings with different thicknesses by the XRD-based $\sin^2\psi$ method [37, 38]. Using the same method, however, Yang et al. [8] experimentally found that the less dense (porosity ~ 10 vol%) HAp coating showed a compressive residual stress of ~5 MPa, while the relatively denser HAp coating (porosity ~ 4 vol%) showed a higher magnitude of compressive residual stress, e.g., ~17 MPa, which was three times higher than that measured for the less dense HAp coating. The presence of more pores and microcracks in the low-density coating was thought [8] to have relaxed the strain and reduced the residual stress.

On the other hand, Ravaglioli and Krajewski [39] reported that the thermal expansion coefficient (CTE, α) of sintered HAp was higher than that of the Ti-6Al-4V alloy ($\alpha_{HAp} = 11.5 \times 10^{-6}$ K^{-1} and $\alpha_{Ti-6Al-4V} = 8.9 \times 10^{-6}$ K^{-1}), and hence the HAp coating might be in a tensile stress state after cooling down from high temperature. However, this suggestion remains debatable. The thermal expansion coefficient is a function of material structure; i.e., in reality, the actual CTE of the HAp coating might be much lower than the theoretical value due to the much lower density of the plasma sprayed HAp coatings. Thus, even for a given method, there exist a lot of variations in both the nature and the magnitude of literature data on residual stress in HAp coatings.

9.6.2 Hole Drilling Method

This technique conforms to the ASTM E837 standard. It is a well-known technique to evaluate residual stress of bulk materials rather than of coatings or thin films. It involves the use of strain gauge at the vicinity of microholes drilled in a given structure. Actually, this process involves drilling a microhole into a residually stressed specimen with a depth that is about equal to its diameter and small compared to the thickness of the test object. That is why this method is identified as semidestructive, because the damage, i.e., the size of the drilled microhole, is really very tiny in size in comparison to that of the sample. In this technique, strain is measured using either a rosette of strain gauges, moire interferometry, laser interferometry based on a rosette of indentations, or holography. The experimental condition should strictly abide by the overall satisfaction of the two basic assumptions: (1) specimen material should be isotropic linearly elastic and (2) in-plane stress gradients should be small.

Han et al. [9] adopted this method and measured by the hole drilling method a maximum tensile residual stress of ~88 MPa for a plasma sprayed HAp coating. However, there had not been many other attempts to utilize this very versatile technique for evaluation of the residual stress in the cases of the HAp coatings.

9.6.3 Raman Piezospectroscopy-Based Method

Again, the Raman piezospectroscopy-based method is not popular, like the previous one for measuring residual stress of HAp coatings. However, Sergo et al. [6] used the Raman piezospectroscopy analysis to measure the residual stress of APS and vacuum plasma sprayed (VPS) HAp coating. The results reported by them have shown that ~100 MPa tensile residual stresses was present in the APS-HAp coating, and ~60 MPa compressive residual stress was present in the VPS-HAp coating. It is to be noted that in the former case, argon was the primary gas, while a mixture of argon and hydrogen was the primary gas in the latter. However, it was not explained why the nature of the residual stress changed from a tensile state to a compressive state by the variation of plasma processing parameters. It was suggested [6] that the compressive residual stress possibly helped to heal the "mud cracking" of the typical plasma sprayed HAp coatings.

9.6.4 Curvature Method

This method is frequently used to determine the stresses within coatings and films. The deposition of coatings or films can induce stresses that can and do cause the substrate to curve in either a concave or convex manner. Then the aforesaid curvatures can be measured using a direct contact methods like profilometry, strain gauges, etc., or indirect contact methods like video, laser scanning, grids, double-crystal diffraction topology, etc. Finally, the curvature is related to the residual stress using the well-known Stoney's relation.

Yang [40] adopted this method and measured strain in the MAPS-HAp coating. A high-resolution camera has been used to measure the deflection of the metallic pure Ti substrate. In this case, the HAp coating was concave in nature.

9.6.5 Nanoindentation

It is well known that the nanoindentation technique is well established for the evaluation of nanomechanical properties at the local microstructural length scale of the materials. Now, Suresh and Giannakopoulos [31] gave us a useful idea to measure the residual stress by instrumented depth-sensitive indentation or nanoindentation technique. They [31] opined that the presence of a residual compressive stress will oppose the penetration of the indenter into the coating, thus resulting in a lower residual depth of penetration in comparison to what would be obtained in a stress-free coating. Similarly, the presence of a residual tensile stress will aid the penetration of the indenter into the coating, thereby leading to a higher residual depth of penetration in comparison to that of a stress-free coating.

The schematics of the nanoindentation processes on a stress-free coating without substrate, a coating with residual compressive stress, and a coating with residual tensile stress along with substrate are shown in Figure 9.2a–c.

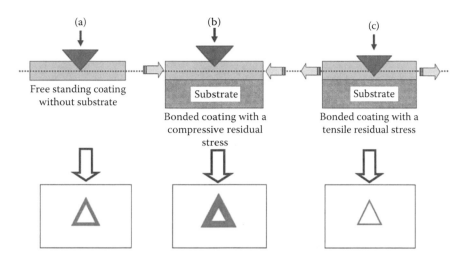

FIGURE 9.2 (See color insert.)
Schematic of nanoindentation processes on a (a) freestanding coating without substrate and bonded coating with a (b) compressive and (c) tensile stress.

Thus, for given fixed load, the residual depth of penetration, and hence the projected contact area, shall be lower in a coating on a given substrate with residual compressive stress (**Figure 9.2b**) in comparison to those of a stress-free coating (**Figure 9.2a**) without the substrate. Similarly, the residual depth of penetration (for the same given fixed load), and hence the projected contact area, shall be higher in a coating (on the same given substrate) with residual tensile stress (**Figure 9.2c**) in comparison to those of a stress-free coating (**Figure 9.2a**) without the substrate. Further, the efficacy of the aforesaid technique has been successfully verified experimentally by the present authors [36].

The ratio of the projected contact area is related to residual tensile stress (σ_t) present in a coating by the following relationship [31]:

$$\frac{A_t}{A_0} = \frac{h_t}{h_0} = \left(1 - \frac{\sigma_t \cdot \sin\beta}{H_t}\right) \tag{9.8}$$

Similarly, the ratio of the projected area is linked to the residual compressive stress (σ_c) present in a coating by the following relationship [31]:

$$\frac{A_c}{A_0} = \frac{h_c}{h_0} = \left(1 + \frac{\sigma_c \cdot \sin\beta}{H_c}\right)^{-1} \tag{9.9}$$

A is the projected area of the bonded coating with residual stress (tensile/compressive). The term A_0 represents the projected area of freestanding coating without the substrate. The quantity h stands for the final depth of penetration of the bonded coating with residual stress (tensile/compressive). Similarly, h_0 is the final depth of penetration of freestanding coating without substrate. H is hardness of the coating, and β is the angle complementary to the semiapex angle of the indenter (i.e., $\beta = (90° − 65.3°) = 24.7°$). The subscripts t and c stand for tensile and compressive nature of residual stress, respectively.

9.6.6 Analytical Models

Besides the direct measurement techniques, the residual stress of plasma sprayed HAp coatings on titanium substrates was studied by Tsui et al. [7] through an analytical model that estimated tensile residual stresses of 21–41 MPa at the top surface of the HAp coatings. However, it would appear from the survey of literature data that there had not been many other attempts to utilize this very versatile technique for evaluation of the residual stress in HAp coatings.

9.7 Role of Higher Plasmatron Power and Secondary Gas

Tsui et al. [7] also opined that using a higher power or secondary gas would promote a residual compressive stress to occur in the plasma sprayed HAp coating. Furthermore, the plasma sprayed specimen of HAp on titanium exhibited a convex curvature that provided a clue suggesting that the residual stress was in fact compressive [21].

9.8 Role of the Substrate Temperature

As mentioned earlier, Yang and Chang [4] produced six plasma sprayed HAp coatings on Ti-6Al-4V substrates of different initial temperatures with various cooling media used during spraying. The residual stresses of six coatings were measured by the XRD-$\sin^2\psi$ method, as discussed earlier.

The HAp coatings with the initial substrate temperatures of 250°C before plasma spraying had the highest compressive residual stress of about 22 MPa. This was suggested [4] to be due to the higher thermal expansion mismatch between coating and substrate during cooling from an elevated temperature to room temperature.

In contrast, for the initial substrate temperature of 25°C, the lowest residual compressive stress of about 17 MPa was obtained [4] due to a lower thermal expansion mismatch between the substrate and the HAp coating. However, for the initial substrate temperature of 160°C, a compressive residual stress of ~20 MPa was recorded [4]. This value was intermediate to those of the two HAp coatings reported above.

In the other case, Yang and Chang [12] measured the residual strain and stress of plasma sprayed HAp coatings on metallic titanium by the materials removal method, which was similar to that employed by Hobbs and Reiter [41]. In addition, the residual stress values of the three HAp coatings measured by the XRD method were quite consistent with those calculated from the materials removal method [12]. The magnitudes of the compressive residual stresses obtained from both measurement methods were found to be linearly dependent on the deposition temperatures of the HAp coatings. Thus, in effect, the variations in the coating temperatures, caused by variations of the cooling media during plasma spraying, were found to have a significant effect on the residual stress states of the resultant HAp coatings.

9.9 Nature of the Residual Stress State

As mentioned earlier, the nature of the residual stress state in the HAp coatings is yet to be unequivocally established. For example, Millet et al. [11], Heimann et al. [42], Sergo et al. [6], and Yang and Chang [4, 8, 12] mentioned that a compressive residual stress was found for HAp coatings on titanium alloy. However, according to the findings of Tsui et al. [7], Tadano et al. [43], and Han et al. [9], a state of tensile residual stresses prevailed in the HAp coatings. These apparent contradictions are yet to be resolved. Hence, more basic studies are needed to explore in more detail the residual stress states in the plasma sprayed HAp coatings.

9.10 Role of Other Basic Process Parameters

The generation of residual stress is mainly controlled by the substrate temperature during plasma spraying. The substrate temperature is in turn determined by the surface speed of the gun traverse during plasma spraying, the thickness of coating developed per deposition pass, and the cooling efficiency during the plasma processing. Together, these factors may cause a change in fracture mode of the coating system and the bonding strength of the coating.

9.11 Residual Stress of Thermally-Sprayed and Sol-Gel-Derived HAp Coatings

Now we shall concentrate on the information collated in Table 9.1, which is a typical survey of reported data on residual stress of HAp coatings deposited by macroplasma spraying (MAPS), MIPS, HVOF, and sol-gel techniques, and subsequently measured by a variety of methods, like XRD-based techniques, the materials removal method, nanoindentation techniques, and the analytical approach. It is interesting to note from the literature data collation presented in Table 9.1 that the numerical magnitude of residual stress of HAp coatings may vary from as low as 5 MPa to as high as 450 MPa and could be either compressive or tensile in nature.

For HAp coatings on Ti-6Al-4V, application of Hooke's law showed that [5] the coatings prepared by the HVOF process had a tensile residual stress of magnitude (e.g., 70–120 MPa) lower than that (e.g., 200–450 MPa) of the coatings prepared by the APS process. Further, the thicker the coating, the higher the magnitude of tensile residual stress estimated. To the contrary, much lower magnitudes of tensile residual stress, e.g., 20–30 MPa, were reported [10] for HAp coatings on Ti-6Al-4V prepared by MAPS in a vacuum environment and detonation gun spraying techniques.

On the other hand, application of the XRD-based $\sin^2\psi$ method gave compressive residual stresses of ~10–30 MPa in HAp coatings on Ti-6Al-4V of different thicknesses [11]. Interestingly, the more dense (e.g., porosity of ~4 vol%) HAp coatings on Ti-6Al-4V showed higher magnitudes of compressive residual stresses (e.g., 17 MPa) than those (e.g., 5 MPa) of the less dense (porosity ~ 10 vol%) coating [8].

Further, adaptation of the same method [11] for MAPS-HAp coatings on Ti-6Al-4V showed that [3, 4] the magnitude (e.g., ~16–27 MPa) of compressive residual stress may vary depending on the initial temperature (e.g., 25–500°C) and the rate of cooling (e.g., variation in nature and composition of cooling media used during spraying of HAp powder) of the substrate.

In addition, the utilization of the Raman piezospectroscopy technique paradoxically gave a residual stress of either a compressive (~60 MPa) or tensile (100 MPa) nature, depending on whether the plasma spraying of HAp powder was done in vacuum or in air [6]. Interestingly, the analytical models predicted [7] a tensile residual stress of ~20–40 MPa for HAp coating on Ti, while the same was measured by the hole drilling method for a MAPS-HAp coating on Ti-6Al-4V to be two to four times as high, e.g., ~ 90 MPa [9].

On the other hand, the materials removal method [12] showed that compressive residual stress was ~30–53 MPa on the top surface of the coating, increasing further (e.g., 48–78 MPa) toward the coating/substrate (e.g., HAp/Ti-6Al-4V) interface. Recently, for fluoridated HAp/β-TCP composite coating

TABLE 9.1

Literature Status on Residual Stress of HAp Coatings

Coating/Substrate	Processing Route	Method of Measurement	Residual Stress (MPa)	Reference
F-HAp, β-TCP/Ti-6Al-4V	SG	XRD ($\sin^2\psi$, $\psi = 90°$)	79–286	[2]
F-HAp, β-TCP/Ti-6Al-4V	SG	$\sigma = (\Delta\alpha \times \Delta t \times E)/(1-\nu)$	50	[2]
HAp/Ti-6Al-4V	MAPS	XRD ($\sin^2\psi$)	(−16) – (−18)	[3]
HAp/Ti-6Al-4V	MAPS	XRD ($\sin^2\psi$)	(−17) – (−27)	[4]
HAp/Ti-6Al-4V	MAPS	Hooke's law + XRD ($\sigma = E\varepsilon$)	200–450	[5]
HAp/Ti-6Al-4V	HVOF	Hooke's law + XRD ($\sigma = E\varepsilon$)	70–110	[5]
HAp/Ti-6Al-4V	MAPS	Raman piezospectroscopy	(−60) – 100	[6]
HAp/Ti	MAPS	Analytical model	21–41	[7]
HAp/Ti-6Al-4V	MAPS	XRD ($\sin^2\psi$)	(−5) – (−17)	[8]
HAp/Ti-6Al-4V	MAPS	Hole drilling	88	[9]
HAp/Ti-6Al-4V	MAPS*	Hooke's law + XRD ($\sigma = E\varepsilon$)	21	[10]
HAp/Ti-6Al-4V	MAPS	Hooke's law + XRD ($\sigma = E\varepsilon$)	29	[10]
HAp/Ti-6Al-4V	MAPS	XRD ($\sin^2\psi$)	(−12.5) – (−29)	[11]
HAp/Ti-6Al-4V	MAPS	XRD ($\sin^2\psi$)	(−36) – (−78)	[12]
HAp/Ti-6Al-4V	MAPS	Materials removal	(−36) – (−53)	[12]
HAp/SS316L	MIPS	Nanoindentation Hooke's law + XRD ($\sigma = E\varepsilon$)	(−22) (−11)	[36]
HAp/Ti-6Al-4V	MIPS	Nanoindentation Hooke's law + XRD ($\sigma = E\varepsilon$)	11 12	[36]

Source: Reprinted/modified from Dey and Mukhopadhyay, *Ceramics International*, 40A: 1263–1272, 2014. With permission from Elsevier.

Note: MAPS = macroplasma spraying technique, MIPS = microplasma spraying technique, HVOF = high-velocity oxy fuel spraying technique, LD = laser-assisted deposition technique, SG = sol-gel technique, F-HAp = fluoridated HAp.

[a] In vacuum.

on Ti-6Al-4V substrate, Cheng et al. [2] estimated the tensile residual stress of ~80–300 MPa by the XRD technique, while the same was estimated as 50 MPa on the basis of difference in coefficient of thermal expansion.

Thus, from the above discussions on pertinent reported literature (Table 9.1), it is noticeable that in spite of the wealth of literature, the data on the nature and magnitude of residual stress were not always so consistent. These apparent contradictions are yet to be resolved. Hence, more basic studies are needed to explore in more detail the residual stress state in the plasma sprayed HAp coatings.

It is important to further note in this context that the present authors have demonstrated [36] that the residual stress for MIPS-HAp coatings can be measured by utilizing a novel nanoindentation technique.

9.12 Residual Stress of MIPS-HAp Coatings

In earlier chapters, we discussed individual advantages and disadvantages of MAPS and MIPS processed HAp coatings. We have seen that the crystallinity of the MIPS-HAp coatings is much higher than that of the MAPS-HAp coatings [44, 45]. Actually, possibly it happens due to lesser plasmatron power inputs of the microplasma spray process, which can reduce the chance of overheating of the molten splats that often pilots to amorphous or glassy phase formation in the HAp coatings prepared by the conventional high-power macroplasma spray technique. Due to the less amorphous phase, it can cause a corresponding enhancement of degree of crystallization of the HAp coatings developed by the MIPS technique. This improvement in crystallinity would undoubtedly assist to diminish the magnitude of residual stress. It will happen simply because the more properly the crystals are grown, the quicker the misfit strains shall reduce. Furthermore, it is already well documented that the porosity level of MIPS-deposited HAp coatings is much higher than that of the MAPS-HAp coatings [44, 45]. These porosities in microstructure may further be active as a stress arrester, or in other words, act as a cushion to accommodate the misfit strains and in that way reduce the magnitude of residual stress.

The residual stress of the as-sprayed MIPS-HAp coating was measured by the nanoindentation technique for the first time ever by the present authors [36]. This technique was originally developed for application to homogeneous, dense material, and its application in the present work requires some clarification, explained above. By an extension of this method to the heterogeneous MIPS-HAp coating, it was initially assumed that within the very small spatial volume that the nanoindenter penetrates at an applied load of 10 mN, the coating is homogeneous. Another important point to be noted here is that nanoindentation measurements were done on the plan section of the MIPS-HAp coating. Thus, the residual stress estimated here refers to the in-plane condition rather than the out-of-plane condition. Further, in the strictest sense, the residual stress measured by the nanoindentation technique can also be considered as a local signature rather than as a global signature. In fact, the nanoindentation technique is possibly one unique method that can, in some way, map the local variation of residual stress in the microstructure.

In the very recent work of Dey and Mukhopadhyay [36] it has been reported that for the MIPS-HAp coatings on SS316L and Ti-6Al-4V, the compressive residual stress of magnitude ~21.74 MPa and tensile residual stress of magnitude ~10.75 MPa, respectively, were measured by the nanoindentation technique (Table 3).

The corresponding projected area of contact (A_0) and the corresponding nanohardness (H) of the freestanding HAp coating were 8.40175×10^5 nm^2

and 4.31 GPa, respectively. The freestanding coating or substrate-free HAp coatings were obtained by the substrate removal process by the surface grinding technique.

It may be noted further that MIPS-HAp coatings on SS316L showed a projected area of contact (A_c) of ~8.22443 × 10^5 nm^2, which was marginally less in comparison to that (A_0) of the freestanding HAp coating [36]. The corresponding nanohardness was also improved to ~4.96 GPa, as A_c was inferior to A_0 of the freestanding HAp coating. This behavior is expected according to Suresh and Giannakopoulos's hypothesis [20].

On the other hand, an opposite trend has been revealed for MIPS-HAp coatings on Ti-6Al-4V. Here it showed a comparatively higher projected area of contact (A_t), e.g., ~8.41180 × 10^5 nm^2, which was superior to that (A_0) of the freestanding HAp coating. Accordingly, as expected [20], the corresponding nanohardness of the MIPS-HAp coatings on Ti-6Al-4V was also decreased to ~3.77 GPa. According to Suresh and Giannakopoulos [20], the experimentally observed decrease of projected area of contact in the MIPS-HAp coating on the SS316L sample was a mark of the existence of a residual compressive stress nature in the surface. Likewise, the experimentally observed increase in the projected area of contact in the MIPS-HAp coating on the Ti-6Al-4V sample suggested the presence of a residual tensile stress state in the surface.

Dey and Mukhopadhyay [36] have also established the efficacy of the nanoindentation technique by reevaluating the residual stress data by the conventional XRD method utilizing Hooke's law, and then comparing the two evaluated data sets on residual stress magnitude and nature. For freestanding HAp coating, the value of E_0 was measured as 30 GPa [36]. The strain (ε) values were calculated as $[(d_c - d_0)/d_0]$, where (d_c) is the interplanner spacing of the coating and (d_0) is the interplanner spacing of the powder according to Equation (9.2). For MIPS-HAp coatings on SS316L substrates, a residual compressive stress of ~11 MPa was evaluated. Likewise, the MIPS-HAp coatings on Ti-6Al-4V substrates showed a residual tensile stress of ~12 MPa. It is further interesting to note that the magnitudes of these residual stress data were nearly similar to and of the order of the residual stress data measured by the nanoindentation technique.

As reported in [36], the d value of the MIPS-HAp coating on SS316L substrate was less than that of the HAp powder (Ca/P-1.667, crystallinity (>80%)), while that of the MIPS-HAp coating on Ti-6Al-4V was superior to that of the same HAp powder. This information explains why the residual stress state was compressive in the MIPS-HAp coating on the SS316L sample, but altered to a tensile one in the MIPS-HAp coating on the Ti-6Al-4V sample.

Dey and Mukhopadhyay [36] also showed that CTE of pure HAp (e.g., $\alpha_{HAp} = 11 \times 10^{-6}$ K^{-1}) was much superior to the CTE value of substrate SS316L (i.e., $\alpha_{SS} = 18 \times 10^{-6}$ K^{-1}). Therefore, the state of residual stress was expected to be compressive in nature, which is indeed experimentally found. It could be argued in an analogous manner [36] that the CTE value of HAp was much less than of the CTE of other substrate Ti-6Al-4V (e.g., $\alpha_{TA} = 9 \times 10^{-6}$ K^{-1}) alloy. Thus, the nature of residual stress was expected to be tensile for MIPS-HAp coating on the Ti-6Al-4V substrate.

It was demonstrated further that the magnitude and nature of residual stress estimated by the nanoindentation method matched well with those evaluated by the well-established XRD-based measurement method. Thus, the efficacy of the nanoindentation technique in evaluation of the residual stress state, as demonstrated in [36], would suggest that the utility of the method should be explored for other ceramic coating or thin-film systems.

Finally, a critical look at the literature data collation presented in Table 9.1 does indeed point out that, barring a few [3, 8, 11] exceptions, in general, the magnitude of residual stress in MIPS-HAp coatings was less than that present in MAPS-HAp coatings. As already discussed above, this can happen as both the degree of crystallinity and volume percent open porosity in MIPS-HAp coatings are higher than those of the MIPS-HAp coatings.

9.13 Summary

In this chapter, the aspect of residual stress of HAp coatings has been highlighted. The origin of residual stress and why it is harmful for in vivo applications have been explained. Several techniques and their merits and demerits for measuring residual stress have been discussed. The residual stresses of HAp coatings prepared by MAPS, MIPS, and other processes have been well documented, while the efficacy and applicability of the newly evolved nanoindentation-based technique were established by examining the residual stress state in MIPS-HAp coatings deposited on Ti-6Al-4V and SS316L substrates. It was further demonstrated that the magnitude and nature of residual stress states estimated by the nanoindentation-based measurement technique matched well with those evaluated by the well-established XRD-based measurement technique.

In vivo studies, i.e., animal trials of microplasma sprayed HAp coatings, will be discussed next in Chapter 10. The in vivo application of MIPS-HAp coatings in rabbit model, goat model, and dog model will be illustrated next. Results obtained from evaluations of serum calcium, inorganic phosphorus and alkaline phosphatase, radiographic, histological, and fluorochrome labeling examinations shall be discussed. Further, the mechanical behavior of MIPS-HAp coatings after in vivo implantation will be described in Chapter 10.

References

1. Matejicek J. and Sampath S. 2001. Intrinsic residual stresses in single splats produced by thermal spray processes. *Acta Materialia*, 49: 1993–1999.
2. Cheng K., Zhang S., Weng W., Khor K. A., Miao S., and Wang Y. 2008. The adhesion strength and residual stress of colloidal-sol gel derived [beta]-tricalcium-phosphate/fluoridated-hydroxyapatite biphasic coatings. *Thin Solid Films*, 516: 3251–3255.
3. Yang Y. C. and Chang E. 2003. The bonding of plasma-sprayed hydroxyapatite coatings to titanium: effect of processing, porosity and residual stress. *Thin Solid Films*, 444: 260–275.
4. Yang Y. C. and Chang E. 2001. Influence of residual stress on bonding strength and fracture of plasma-sprayed hydroxyapatite coatings on Ti-6Al-4V substrate. *Biomaterials*, 22: 1827–1836.
5. Brown S. R., Turner I. G., and Reiter H. 1994. Residual stress measurements in thermal sprayed hydroxyapatite coatings. *Journal of Materials Science: Materials in Medicine*, 5: 756–759.
6. Sergo V., Sbaizero O., and Clarke D. R. 1997. Mechanical and chemical consequences of the residual stresses in plasma sprayed hydroxyapatite coatings. *Biomaterials*, 18: 477–482.
7. Tsui Y. C., Doyle C., and Clyne T. W. 1998. Plasma sprayed hydroxyapatite coatings on titanium substrates part 1: mechanical properties and residual stress levels. *Biomaterials*, 19: 2015–2029.
8. Yang Y. C., Chang E., Hwang B. H., and Lee S. Y. 2000. Biaxial residual stress states of plasma-sprayed hydroxyapatite coatings on titanium alloy substrate. *Biomaterials*, 21: 1327–1337.
9. Han Y., Xu K., and Lu J. 2001. Dissolution response of hydroxyapatite coatings to residual stresses. *Journal of Biomedical Materials Research*, 55: 596–602.
10. Gledhill H. C., Turner I. G., and Doyle C. 2001. In vitro dissolution behaviour of two morphologically different thermally sprayed hydroxyapatite coatings. *Biomaterials*, 22: 695–700.
11. Millet P., Girardin E., Braham C., and Lodini A. 2002. Stress influence on interface in plasma-sprayed hydroxyapatite coatings on titanium alloy. *Journal of Biomedical Materials Research*, 60: 679–684.
12. Yang Y. C. and Chang E. 2005. Measurements of residual stresses in plasma-sprayed hydroxyapatite coatings on titanium alloy. *Surface and Coatings Technology*, 190: 122–131.
13. Ruckle D. L. 1980. Plasma-sprayed ceramic thermal barrier coatings for turbine vane platforms. *Thin Solid Films*, 73: 455–461.
14. Elsing R., Knotek O., and Balting U. 1990. The influence of physical properties and spraying parameters on the creation of residual thermal stresses during the spraying process. *Surface and Coatings Technology*, 41: 147–156.
15. Dally J. W. and Riley W. F. 1978. *Experimental stress analysis*. New York: McGraw-Hill.

16. Scardi P., Leoni M., and Bertamini L. 1996. Residual stresses in plasma sprayed partially stabilised zirconia TBCs: influence of the deposition temperature. *Thin Solid Films*, 278: 96–103.

17. Takeuchi S., Ito M., and Takeda K. 1990. Modelling of residual stress in plasma-sprayed coatings: effect of substrate temperature. *Surface and Coatings Technology*, 43/44: 426–435.

18. Noutomi A., Kodama M., Inoue Y., One T., Kawano M., and Tani N. 1989. Residual stress measurement on plasma sprayed coatings. *Welding International*, 11: 947–953.

19. Eigenmann B., Scholtes B., and Macherauch E. 1989. Determination of residual stresses in ceramics and ceramic-metal composites by x-ray diffraction methods. *Materials Science and Engineering A*, 118: 1–17.

20. Evans A. G., Crumley G. B., and Demaray R. E. 1983. On the mechanical behavior of brittle coatings and layers. *Oxidation of Metals*, 20: 196–216.

21. Mencik J. 1995. *Mechanics of components with treated or coated surfaces*. London: Kluwer Academic Publishers, 202–205.

22. Mevrel R. 1987. Cyclic oxidation of high-temperature alloys. *Materials Science and Technology*, 3: 531–535.

23. Wang B. C., Chang E., Yang C. Y., Tu D., and Tasi C. H. 1993. Characteristics and osteoconduction of three different plasma-sprayed hydroxyapatite-coated titanium implants. *Surface and Coatings Technology*, 58: 107–117.

24. Clyne T. W. and Gill S. C. 1996. Residual stresses in thermal spray coatings and their effect on interfacial adhesion: a review of recent work. *Journal of Thermal Spray Technology*, 5: 401–418.

25. Kesler O., Finot M., Suresh S., and Sampath S. 1997. Determination of processing-induced stresses and properties of layered and graded coatings: experimental method and results for plasma-sprayed Ni-Al$_2$O$_3$. *Acta Materialia*, 45: 3123–3134.

26. Cullity B. D. *Elements of x-ray diffraction*. Reading, MA: Addison-Wesley, 447.

27. Kuroda S., Kukushima T., and Kitahara S. 1988. Simultaneous measurement of coating thickness and deposition stress during thermal spraying. *Thin Solid Films*, 164: 157–163.

28. Gill S. C. and Clyne T. W. 1994. Investigation of residual stress generation during thermal spraying by continuous curvature measurement. *Thin Solid Films*, 250: 172–180.

29. Sue J. A. 1992. X-ray elastic constants and residual stress of textured titanium nitride coating. *Surface and Coatings Technology*, 54/55: 154–159.

30. Perry A. J., Albert S. J., and Martin P. J. 1996. Practical measurement of the residual stress in coatings. *Surface and Coatings Technology*, 81: 17–28.

31. Suresh S. and Giannakopoulos A. E. 1998. A new method for estimating residual stresses by instrumented sharp indentation. *Acta Materialia*, 46: 5755–5767.

32. Bouzakis K. D., Skordaris G., Mirisidis J., Hadjiyiannis S., Anastopoulos J., Michailidis N., Erkens G., and Cremer R. 2003. Determination of coating residual stress alterations demonstrated in the case of annealed films and based on a FEM supported continuous simulation of the nanoindentation. *Surface and Coatings Technology*, 174–175: 487–492.

33. Bouzakis K. D. and Michailidis N. 2004. Coating elastic-plastic properties determined by means of nanoindentations and FEM-supported evaluation algorithms. *Thin Solid Films*, 469/470: 227–232.

34. Taylor C. A., Wayne M. F., and Chiu W. K. S. 2003. Residual stress measurement in thin carbon films by Raman spectroscopy and nanoindentation. *Thin Solid Films*, 429: 190–200.

35. Chudoba T., Schwarzer N., Linss V., and Richter F. 2004. Determination of mechanical properties of graded coatings using nanoindentation. *Thin Solid Films*, 469/470: 239–247.

36. Dey A. and Mukhopadhyay A. K., 2014. Evaluation of residual stress in microplasma sprayed hydroxyapatite coating by nanoindentation. *Ceramics International*, 40A: 1263–1272.

37. Perry A. J., Albert S. J., and Martin P. J. 1996. Practical measurement of the residual stress in coatings. *Surface and Coatings Technology*, 81: 17–28.

38. Noyan I. C. and Cohen J. B. 1987. *Residual stress*. New York: Springer.

39. Ravaglioli A. and Krajewski A. 1992. *Bioceramics: materials, properties, application*. London: Chapman & Hall, 44–45.

40. Yang Y. C. 2007. Influence of residual stress on bonding strength of the plasma-sprayed hydroxyapatite coating after the vacuum heat treatment. *Surface and Coatings Technology*, 201: 7187–7193.

41. Hobbs M. K. and Reiter H. 1988. Residual stresses in ZrO_2–8%Y_2O_3 plasma-sprayed thermal barrier coatings. *Surface and Coatings Technology*, 34: 33–42.

42. Heimann R. B., Grassmann O., Hempel M., Bucher R., and Harting M. 2000. Phase content, resorption resistance and residual stresses in bioceramic coatings. In *Applied mineralogy*, Rammlmair D., Mederer F., Oberthur T. H., Heimann R. B., and Pentinghaus H. (Eds.), vol. 1. Brookfield, 155.

43. Tadano S., Todoh M., Shibano J., and Ukai T. 1997. Influence of residual stress on bonding strength of plasma-sprayed hydroxyapatite coatings on Ti6Al4V substrate. *SME International Journal Series A: Mechanical and Materials Engineering*, 40: 328.

44. Dey A., Mukhopadhyay A. K., Gangadharan S., Sinha M. K., and Basu D. 2009. Development of hydroxyapatite coating by microplasma spraying. *Materials and Manufacturing Processes*, 24: 1321–1330.

45. Dey A., Mukhopadhyay A. K., Gangadharan S., Sinha M. K., and Basu D. 2009. Characterization of microplasma sprayed hydroxyapatite coating. *Journal of Thermal Spray Technology*, 18: 578–592.

10

In Vivo Studies of Microplasma Sprayed Hydroxyapatite Coating

10.1 Introduction

This chapter will tell us the about the real-life implantation possibility of the MIPS-HAp coating as a new genre of bioactive material. Here, we will discuss the MIPS-HAp-coated Ti and SS3161 implants in the rabbit model, goat model, and the dog model. We will also emphasize the biochemical, histological, radiographic, and fluorochrome labeling studies on MIPS-HAp-coated intramedullary metallic, i.e., SS3161, pins for fracture repair of surgically created artificial defects in the tibia of New Zealand white rabbits, in particular. The idea behind presentation of these exciting new results is to create further motivation for future researchers to advance the MIPS-HAp coating technology toward the human trial in days to come following, of course, all the mandatory ethical as well as technological issues that may need to be overcome within a reasonable time frame and with resources as deemed appropriate.

10.2 Rabbit Model

10.2.1 Intramedullary Pinning

Before we can go into the details of intramedullary pinning, we need to know what it actually means. Well, simplistically speaking, it is one of the most commonly and conveniently used fracture fixation techniques. The basic idea that works behind this technique is provision of stability to long bone fractures, especially in animals [1, 2]. One of the major advantages of such a pinning procedure is that such implants can be removed with relative ease following the time when the fractured zone has healed up. However, as the saying goes, life is not always full of roses. It cannot be. Therefore, this simple pinning technique comes not only with its gains or advantages, but also the disadvantages or demerits of the same. One major disadvantage is that in this case, the stability of the fracture-managed part improves, but only slowly. As a result, the return-back path time to the normal function of the affected limb also becomes a little bit longer than what would have made us happy. A consequence of these two steps is that post-operative care following this type of pinning has to be much more than what is typically required for other, more conventional types of internal fixation techniques. There can be delayed union or even nonunion of the separated parts of the bone. This, of course, would be a rare situation. The other such extreme situation may, at least in principle, involve excessive flexibility, pin deformation, fatigue fracture, or even migration of the pin. To overcome such situations, basic research in the fracture healing is essential to determine the optimum quality of materials, as well as the geometry and structural properties of fixation devices. From the material science point of view, the most lovable or ideal situation will certainly be the development of such a composite material that can mimic the natural bone, both composition-wise and mechanical property-wise.

Is there any material that can exactly mimic bone? The answer to this question may not be affirmative as of now. Nevertheless, a large number of materials do exist that in some way or other either mimic a part of a bone or provide some extraordinary technological or biological advantage. Thus, a host of materials have the capability to fill and reconstruct bone defects. At least in principle, the list may include a variety as wide as, but not necessarily limited to, polymers, bioactive glasses and bioceramics, collagen, metals, metal alloys, carbon-based materials, and possibly composite formulation of these materials. For instance, the bioactive ceramics have two unique advantages. They can act both osteoinductively and osteoconductively. Another uniqueness of bioactive ceramics is that they can also act as a scaffold. Such a material can and really does enhance bone formation on its surface. Therefore, it is possible to take advantage of this scenario. Such bioactive ceramics can then be conveniently used as a coating on various substances. Also, they could be utilized in porous form to fill bone defects.

Thus, it is only natural that during the past two decades there has been almost a dramatic increase in the research and development on bioceramic materials. It is already well known that the primary inorganic component of bone is hydroxyapatite (HAp). In turn, HAp belongs to the calcium phosphate group. Moreover, it is further known that calcium phosphates are osteoconductive, osteointegrative, and in a few cases, osteoinductive. That is why calcium phosphate HAp, etc., bioceramics are most useful for the cortical part of the bone structure. Whenever an internal fixation is achieved through the use of metallic biomedical implants, there is more often than not a chance of implant loosening/migration, as mentioned many a times earlier in the chapters of this book. A completely satisfactory solution to these problems is yet to emerge. In other words, implant performance failure due to loosening or migration is an unresolved issue that deserves the highest possible attention from both perspectives of patient satisfaction and patient comfort. The genesis of this problem lies at the bone-implant interface.

Therefore, to overcome this problem, the bone-implant interface is modified by using especially hydroxyapatite, β-tricalcium phosphate and its composite coatings to provide improved osseointegration [3–6]. Thus, as mentioned earlier, the coating of intramedullary metallic pins with HAp, tricalcium phosphate (TCP), etc., is a commonly used fracture fixation technique to provide stability to long bone fractures [1, 2, 7–10]. Their choice is guided by the fact that they can provide easier ingrowth of bony tissues [11, 12].

Junker et al. [13–15] recently made a very significant effort in bone fracture fixation. They have used microplasma sprayed (MIPS) calcium phosphate-coated Ti implants in goat and dog. However, to the best of our knowledge, the work reported by Dey and coworkers [16], as described in this book, is the first comprehensive report on the comparative healing performance efficacies of uncoated intramedullary SS316L pins and MIPS-pure HAp-coated intramedullary SS316L pins. Thus, this work in reference has attempted to fix bone defects in New Zealand white rabbits using the aforesaid intramedullary pinning arrangements.

Surgical-grade SS316L intramedullary pins were coated by the optimized MIPS technique [16]. Prior to in vivo animal trials, the actual uncoated pins and pins coated by MIPS-HAp are shown in Figure 10.1a and b, respectively. The present study was carried out following the procedures conforming to the standards of the Institutions Animal Ethical Committee of the West Bengal University of Animal and Fishery Sciences, Kolkata, India. In this present in vivo investigation, 18 healthy and adult New Zealand white rabbits

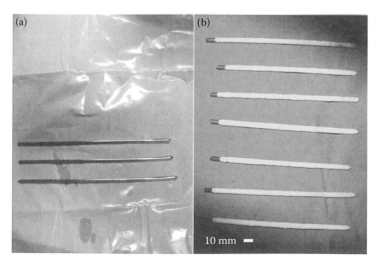

FIGURE 10.1 (See color insert.)
SS316L intramedullary pins: (a) uncoated and (b) MIPS-HAp coated.

FIGURE 10.2 (See color insert.)
Progressive steps (marked as arrows from right to left) of implantation of MIPS-HAp-coated intramedullary pin into the New Zealand white rabbit: deliberate creation of defect by micro-drilling, closing muscle layers, and closure out of the skin.

of 12–15 months age and either sex and weighing ~2.5 kg were utilized. For the present rabbit model, the following protocol was used:

1. Routine preoperative preparation.
2. Rabbits were anaesthetized using a combination of xylazine hydrochloride and ketamine hydrochloride anesthesia. This standard process was described by Sedgwick [185] earlier.
3. Shaving of the hair.
4. A rectangular tiny defect was deliberately created in the cortical bone by an electrically operated micromotor-based dental drill.
5. Then intramedullary pins were implanted within the medullary cavity of the tibia of rabbits as per the procedure reported by DeYoung and Probst [1] and Olmstead et al. [2].
6. Chromic catgut (3/0) was utilized for closing incisions, and skin of the rabbits was sutured with silk (3/0) as per routine surgical protocol.
7. The wound was covered with a protective dressing and bandages. Routine dressing of the wound was done on the 3rd day onward, on alternative days up to the 7th day post-operatively; subsequently, the skin sutures were eradicated on the 10th day after the operation. The steps of the surgery processes are shown in Figure 10.2, depicting defect creation, insertion of the metallic pins, and closing of the skin by stitching.

10.2.2 Visual Observation

Bone defect or fracture repair ability and the signs of local inflammatory reactions, if any, were observed up to the 60th post-operative day. Mild swelling was observed at the surgery site. However, by the 10th post-operative day, it subsided gradually and returned to normalcy. In the cases of both MIPS-HAp-coated implants and bare uncoated implants, there were exhibitions of apparent lameness in the immediate post-operative period, as expected. The lameness, as well as transient inflammatory response observed in the immediate post-operative period, can be plausibly argued to be due to surgical and operative trauma. Further, as post-operative days gradually passed one after another, the lameness disappeared. This information indicates that the inflammation had subsided and the defect/fracture site had become stable. These observations were similar to those reported in the cases of the rabbit model and dog model [17, 18]. Now, we want to highlight another important finding. It is to be emphasized that in the present study, no toxicity was elicited. It can be inferred from this that the inserts were well accepted by the body without any undesirable effect.

10.2.3 Studies of Serum Calcium, Inorganic Phosphorus, and Alkaline Phosphatase

10.2.3.1 Serum Calcium

The amounts of serum calcium, inorganic phosphorus, and alkaline phosphatase were evaluated in the present study on operative day 0 to post-operative day 60 to investigate the degree of inflammation and progress of bone healing or reconstruction. The experiments were conducted following [19–21].

Changes in serum alkaline phosphatase (ALP) activity (IU/L) at different durations of fracture healing for animals inserted with uncoated and MIPS-HAp-coated pins are shown in Figure 10.3a. The present study showed significant changes in serum alkaline phosphatase activity (IU/L) at different durations of bone fracture healing in rabbits inserted with both uncoated and MIPS-HAp-coated pins. A look at these data confirms that the serum ALP level increased significantly on the 15th post-operative day, while it decreased gradually to normal after the 60th post-operative day. This happens because the osteoblast heals through osteogenesis [22] after secreting alkaline phosphatase enzyme [23]. Further, the extent of rise of serum ALP happened due to the different rates of osteogenesis as per quantitative and qualitative activity of osteoprogenitor cells. Thus, there was initially an enhanced activity of osteoblast. It had led to increased ALP activity up to the 15th post-operative day. Following this period, slowly afterwards the ALP activity gradually returned to normal when callus formation had ceased [24].

10.2.3.2 Inorganic Phosphorus

The changes in serum calcium estimation (mg/dl) at different durations of fracture healing for rabbits inserted with uncoated and MIPS-HAp-coated pins are shown in Figure 10.3b. The data showed no significant changes in serum calcium level (mg/dl) throughout the study period. These data would strongly suggest that fracture healing had nothing to do with serum calcium level, as also opined by Speed [25] and Pandey and Udupa [26]. The present observations, however, were different from those reported by Suryawanshi et al. [27], who found an increase in serum calcium level after fracture.

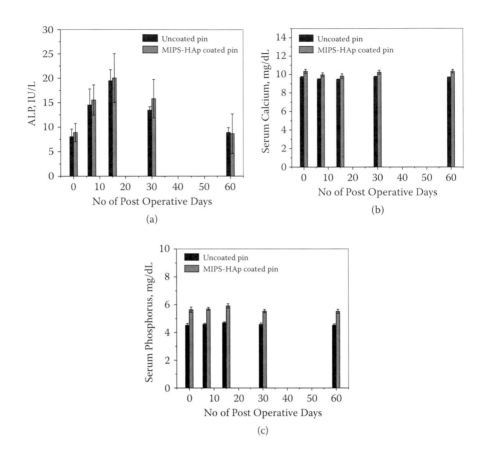

FIGURE 10.3 (See color insert.)
Changes in (a) serum alkaline phosphatase activity (IU/L), serum calcium estimation (mg/dl), and serum phosphorous level (mg/dl) at different durations of fracture healing for animals inserted with uncoated and MIPS-HAp-coated pins. (Reprinted/modified from Dey et al., *Ceramics International*, 8: 1377–1391, 2011. With permission from Elsevier.)

Nevertheless, it was found that up to the 15th post-operative day, in all the rabbits there was a insignificant reduction in serum calcium values. It may be plausibly argued that this could be due to increased urinary excretion after injury that was definitely a traumatic experience for the bone [27].

10.2.3.3 Alkaline Phosphatase

Changes in serum phosphorous level (mg/dl) at different durations of fracture healing for animals inserted with uncoated and MIPS-HAp-coated pins are shown in Figure 10.3c. Like serum calcium estimation, here also the data on the serum phosphorus level (mg/dl) showed no significant changes with number of post-operative days passed during the process of fracture healing.

Other researchers, however, have reported an increase in serum phosphorus level during early stages of fracture healing due to enhanced mineralization [26–28], as well as necrotic disintegration of the cells at the fracture site [27].

10.2.4 Radiographic Evaluation

Radiographs (anterioposterior (A/P)) were studied before, immediately on, and after the implantation to understand the status of implant, soft tissue response, and formation of new bone. Radiographic observation at the fracture site, taken at the 0 and 60th days post-operatively for the MIPS-HAp-coated pins, are shown in Figure 10.4a and b, respectively. A radiolucent rectangular cortical defect on the proximal third of the tibia was seen on the day of operation, i.e., day 0. A uniformly dense margin of the pin margin, along with details, was distinctly observable. However, bony defect was not visible on radiograph on the 60th post-operative day. Further, a better organized and compact appearance of the newly formed osseous tissue was noticeable.

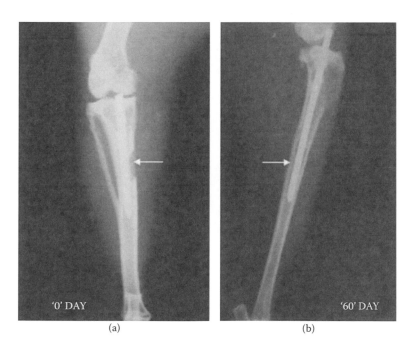

(a) (b)

FIGURE 10.4 (See color insert.)
Radiographic observation at the fracture site, taken at (a) 0 and (b) 60 days post-operatively for the MIPS-HAp-coated pins. The arrow in (a) indicates the artificial defects created in bone, and the arrow in (b) indicates the complete annihilation of the bone defect by newly formed osseous tissue, and thus established cortical continuity. (Reprinted/modified from Dey et al., *Ceramics International*, 8: 1377–1391, 2011. With permission from Elsevier.)

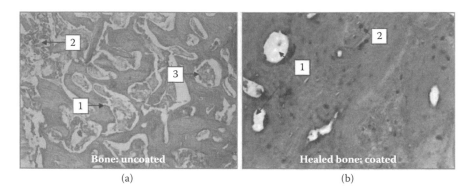

Histopathology photomicrographs of (a) uncoated and (b) MIPS-HAp-coated pins inserted in rabbits. (Reprinted/modified from Dey et al., *Ceramics International*, 8: 1377–1391, 2011. With permission from Elsevier.)

10.2.5 Histological Observations

Bone samples were collected from both defect-induced and normal areas of the animals for histological investigation. This was done on the 60th postoperative day. In a routine manner, the decalcified tissues were processed. The sections cut were about 4 μm in size. These sections were stained with hematoxylin and eosin. This was done to observe the matrix formation and conditions.

Histopathology photomicrographs of uncoated and MIPS-HAp-coated pins inserted in rabbits are shown in Figure 10.5a and b, respectively. The section showed cancellous bone and well-developed marrow spaces in the uncoated group. Osteoblasts had lined the bone lamella. Adipose tissue with marrow material had composed the marrow cavity. This might be due to the fact that the inserted uncoated, i.e., inert intramedullary pins form a fibrous tissue capsule but not a chemical bond with tissues [29]. In contrast, a complete ossification with developed Haversian canals and well-defined peripherally placed osteoblasts was exhibited by the MIPS-HAp-coated group [30–32], as the calcium phosphate coatings had not only improved bone bonding but also promoted more rapid osseointegration. Well-developed blood vessels were found in Haversian spaces. These spaces had very little marrow. The lamellar cortical bone also contained few particles of nonabsorbed biodegradable material. Some refractile crystalloid structures were also seen in the marrow space.

10.2.6 Fluorochrome Labeling Study

Several methods exist to examine newly formed bone using specific bone markers and labeling techniques [33, 34]. The tetracycline labeling method was introduced to measure the quantity of newly formed bone as the tetracycline molecule exhibits a fluorescence property in ultraviolet light. Oxytetracycline follows the ionized calcium and the fixation by a process of adsorption, which is restricted to areas where active deposition of mineralized tissue is taking place [35]. The labeled new bone and old bone emitted bright golden yellow and dark sea green fluorescence, respectively. These parts provide immense value in studying gross bone architecture and histological mapping of new bone formation using fluorescent bone markers. In the present study, oxytetracycline at a dose of 50 mg.kg^{-1} of body weight (2-6-2 pattern) before the end of the study was proved to be sufficient to quantify the extent of new bone formation.

In the present study, fluorochrome was observed by injecting oxytetracycline dehydrate (Pfizer India, India) intramuscularly, at a dose rate of 50 mg/kg of body weight on days 48, 49, and later after 6-day intervals on days 55 and 56 (2-6-2 manner) post-operatively for double toning of new bone. Undecalcified ground sections were prepared from the implanted segments of bone and were ground to 20 µm thickness using different grades of sandpaper. The ground-undecalcified sections were observed under ultraviolet incidental light with an Orthoplan microscope (excitation filter, BP = 400 range; Leitz, USA) for tetracycline labeling to determine the amount and source of newly formed bone.

Fluorochrome photomicrographs of uncoated and MIPS-HAp-coated pins inserted in rabbits are shown in Figure 10.6a, b. In the uncoated group, when viewed under fluorescent light, the ultraviolet light imparted a double-toned golden yellow fluorescence as tetracycline marked the new bone. This golden yellow fluorescence was present as a narrow area in the defect site, whereas the host bone appeared in a dark homogeneous sea-green color. The defect was completely filled with newly formed cancellous bone and appeared as a homogeneous nonfluorescent area. Union in the defect site of the bone was complete in most of the animals. In the MIPS-HAp-coated group, under fluorescent microscope, the defect was visualized as a line of golden yellow

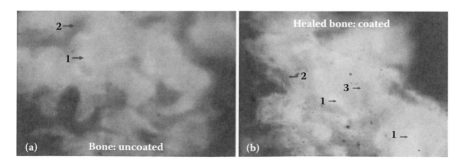

FIGURE 10.6 (See color insert.)
Fluorochrome labeling study: photomicrograph of (a) uncoated and (b) MIPS-HAp-coated pins inserted in rabbits. (Reprinted/modified from Dey et al., *Ceramics International*, 8: 1377–1391, 2011. With permission from Elsevier.)

fluorescence, whereas the host bone appeared as a homogeneous dark sea-green color. In this group, the activity of new bone formation was marked. Within this new osteoid tissue, which completely filled the bone defect, crossing over of new bone trabeculae was evident. Few resorption cavities were also present.

Most of the bone defects were occupied by a homogeneous nonfluorescent area in rabbits of the uncoated group. This observation suggests that in this case, there was very little amount of new bone formation that had happened. In the coated group, however, the golden yellow fluorescence in a narrow area was observed. This observation suggests that new bone formation had taken place in the defect site of the host bone. These observations are similar to those of Singh [18] and Nandi et al. [36, 37]. There was also the additional presence of resorption cavities. This particular observation suggests that the resorption and replacement of the bone were happening simultaneously. In addition, the very presence of the resorption cavities would also indicate the initiation of the most important process of bone remodeling [36, 37] at the defect site.

10.3 Goat Model

In fact, Junker and his coworkers [13] studied in vivo properties in the goat model for the first time for MIPS-coated CaP coatings. Calcium phosphate or CaP bioactive ceramic coatings on screw-type Ti implants had been developed by Junker and his coworkers [13]. Subsequently, they were implanted into trabecular bone of the femoral condyles of female Saanen goats. The ages of the goats were 2–4 years. Their weight was around 60 kg. All prosthesis was sterilized by autoclaving before proceeding for further implantation. Bone fixation properties had been investigated after 6 and 12 weeks after the implantation. They [13] concluded that in repair of bone defects, MIPS CaP-coated Ti implants were as efficient as macroplasma sprayed (MAPS) CaP-coated Ti implants. It is interesting to note that the usage efficacy of conventional MAPS-coated implants for repair of bone defect in goat models has been already well studied by several researchers [38–40].

10.3.1 Visual Observation

All the goats that were implanted had remained in good health after implantation with the MIPS-coated CaP ceramic. To perform other studies, like histological and mechanical properties, the goats were sacrificed at a later stage. But even at that later stage, no signs of inflammation or wound due to implantation surgery were found [13].

10.3.2 Mechanical Behavior

Further, Junker et al. [13] had evaluated the failure load, i.e., bone bonding strength, of the interface between the coating and bone. A specially designed fixture (tensile bench) had been utilized for the same. This special fixture was designed and developed following the work reported by Hulshoff et al. [41]. The specimens in the tensile bench were tested at a constant displacement speed of 0.5 mm.min^{-1}. After 6 weeks, MIPS CaP-coated metallic implants showed better failure strength (e.g., 135–137 N.cm^{-1}) in comparison to the failure strength (e.g., 128 N.cm^{-1}) of the conventional MAPS CaP-coated metallic implants. In contrast, the bare metallic implants showed significantly poor results, e.g., bonding strength of 43 N.cm^{-1}. Further, after 12 weeks of implantation, the failure strength of the MIPS CaP-coated metallic implants showed a minor improvement, and the trend of the data for MAPS- and MIPS-coated implants was similar, as discussed earlier. However, a significant improvement (e.g., bonding strength of 76 N.cm^{-1}) had been seen in the case of the bare metallic implant. The scanning electron microscopy (SEM) photomicrographs of the fractured surfaces for uncoated and MIPS CaP-coated interfaces are shown in Figure 10.7a and b, respectively. Here we can see that the failure took place only at a region of CaP coating and host bone interface, instead of at the coating–metallic implant interface. This information again implies that in terms of coating bonding strength or adhesion property, the reliability of MIPS CaP-coated metallic implants is beyond question and definitely acceptable as a promising area for further research and development.

10.4 Dog Model

Apart from making contributions in the goat model, Junker et al. [14, 15] had also studied the bone formation characteristic in vivo in the dog model. The MIPS–CaP-coated dental implants were utilized for this purpose. They have shown short-term stability of MIPS CaP-coated implants. The MIPS–CaP coatings were deposited on a screw-type Ti dental implant. Subsequently, it was implanted in Beagle dogs 4–5 years in age and with an average weight of ~10 kg. The protocol that had been adopted in this study had been ethically approved by the Committee for Animal Research of the Radboud University Nijmegen Medical Center. For MIPS–CaP-coated Ti implant materials, the response for osseointegration and bone formation was found to be favorable. This information again suggests the promising possibilities of the MIPS-coated metallic implants for human trials in days to come.

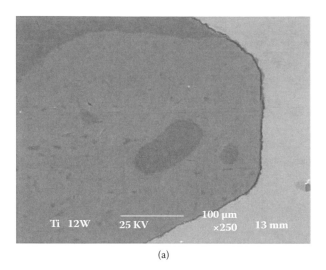

(a)

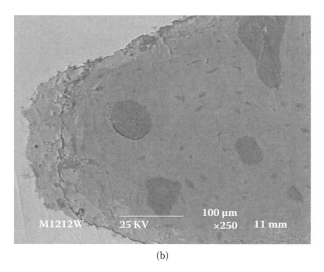

(b)

FIGURE 10.7
Microstructures of (a) uncoated and (b) MIPS-coated CaP bioceramic. (Reprinted/modified from Junker et al., *Clinical Oral Implants Research*, 21: 189–200, 2010. With permission from Wiley.)

Junker and his coworkers [13–15] utilized CaP coatings, which had a comparatively lower crystallinity (e.g., 67–80%) and reported a reduction of coating thickness due to the dissolution of the amorphous phase between the remaining crystalline coating particles of HAp and the presence of other calcium phosphate phases in the coating composition. However, the work of Dey and coworkers [16] utilized highly crystalline (80–91%) and phase-pure HAp.

10.5 Summary

Here we have presented for the very first time a study on the comparative healing performance efficacies of uncoated and MIPS-pure HAp-coated intramedullary SS316L pins for bone defect fixation in New Zealand white rabbits. The results of the biochemical, histological, radiographic, and fluorochrome labeling studies proved more bone bonding and osseointegration in HAp-coated metallic pins leads to complete bone defect healing that is not possible to attain with the uncoated metallic pins. In other words, MIPS-HAp-coated intramedullary SS316L pins have bone healing performance efficacy much superior to that of uncoated metallic pins. The superior efficacy of the MIPS-HAp-coated metallic implants matched the results of additional research conducted by other researchers in goat and dog models. Thus, it may be suggested that MIPS-HAp-coated or CaP-coated metallic implants may have a very promising future for bone defect repair, along with the golden possibility for human trials in days to come.

The future scope and possibilities of plasma spraying, with particular emphasis on MIPS-HAp coatings, will be explored in Chapter 11. It will be shown in Chapter 11 that this technique has the capability to emerge as a potential deposition technique for developments of HAp and HAp composite coatings on complex and contoured implants for human beings. Finally, it will be shown how MIPS-HAp coatings can be deposited on a C/C composite.

Acknowledgments

The contribution of Dr. S. K. Nandi of the West Bengal University of Animal and Fishery Sciences, Kolkata, India, is gratefully acknowledged for the animal trial and providing the pertinent images during insertion of the implants in vivo, shown in Figure 10.2. The authors also acknowledge Dr. B. Kundu of CSIR-CGCRI for the useful interactions and interpretation of the data regarding in vivo animal trial study.

References

1. DeYoung D. J. and Probst C. W. 1993. Methods of internal fracture fixation-general principles. In *Textbook of small animal surgery*, Slatter D. H. and Saunders W. B. (Eds.). Philadelphia: Elsevier Health Sciences, 1610–1631.
2. Olmstead M. L., Egger E. L., and Johnson A. L. 1995. Principles of fracture repair. In *Small animal orthopedics*, Olmstead M. L. (Ed.). Mosby-Year Book, 111–159.

3. Filiaggi M. J., Coombs N. A., and Pilliar R. M. 1991. Characterization of the interface in the plasma-sprayed HA coating/Ti-6Al-4V implant system. *Journal of Biomedical Materials Research*, 25: 1211–1229.
4. Thomas K. A., Kay J. F., Cook S. D., and Jarcho M. 1987. The effect of surface macrotexture and hydroxylapatite coating on the mechanical strengths and histologic profiles of titanium implant materials. *Journal of Biomedical Materials Research*, 21: 1395–1414.
5. Rivero D. P., Fox J., Skipor A. K., Urban R. N., and Galante J. O. 1988. Calcium phosphate-coated porous titanium implants for enhanced skeletal fixation. *Journal of Biomedical Materials Research*, 22: 191–201.
6. Ducheyne P. and Qiu Q. 1999. Bioactive ceramics: the effect of surface reactivity on bone formation and bone cell function. *Biomaterials*, 20: 2287–2303.
7. Lima R. S., Khor K. A., Li H., Cheang P., and Marple B. R. 2005. HVOF spraying of nanostructured hydroxyapatite for biomedical applications. *Materials Science and Engineering A*, 396: 181–187.
8. Gross K. A., Berndt C. C., and Herman H. 1998. Amorphous phase formation in plasma sprayed hydroxyapatite coating. *Journal of Biomedical Materials Research*, 39: 407–414.
9. Tong J., Chen W., Li X., Cao Y., Yang Z., Feng J., and Zhang X. 1996. Effect of particle size on molten states of starting powder and degradation of the relevant plasma-sprayed hydroxyapatite coating. *Biomaterials*, 17: 1507–1513.
10. Rokkum M., Reigstad A., Johansson C. B., and Albrektsson T. 2003. Tissue reactions adjacent to well fixed hydroxyapatite-coated acetabular cups. Histopathology of ten specimens retrieved at reoperation after 0.3 to 5.8 years. *Journal of Bone and Joint Surgery*, 85: 440–447.
11. Ong J. L., Carnes D. L., and Bessho K. 2004. Evaluation of titanium plasma-sprayed and plasma-sprayed hydroxyapatite implants in vivo. *Biomaterials*, 25: 4601–4606.
12. Lee T. M., Wang B. C., Yang Y. C., Chang E., and Yang C. Y. 2001. Comparison of plasma-sprayed hydroxyapatite coatings and hydroxyapatite/tricalcium phosphate composite coatings: in vivo study. *Journal of Biomedical Materials Research*, 55: 360–367.
13. Junker R., Manders P. J. D., Wolke J., Borisov Y., and Jansen J. A. 2010. Bone-supportive behavior of microplasma-sprayed CaP-coated implants: mechanical and histological outcome in the goat. *Clinical Oral Implants Research*, 21: 189–200.
14. Junker R., Manders P. J. D. Wolke J., Borisov Y., and Jansen J. A. 2010. Bone reaction adjacent to microplasma sprayed CaP-coated oral implants subjected to occlusal load, an experimental study in the dog. Part I. Short-term results. *Clinical Oral Implants Research*, 21: 1251–1263.
15. Junker R., Manders P. J. D., Wolke J., Borisov Y., and Jansen J. A. 2011. Bone reaction adjacent to microplasma sprayed calcium phosphate-coated oral implants subjected to an occlusal load, an experimental study in the dog. *Clinical Oral Implants Research*, 22: 135–142.
16. Dey A., Nandi S. K., Kundu B., Kumar C., Mukherjee P., Roy S., Mukhopadhyay A. K., Sinha M. K., and Basu D. 2011. Evaluation of hydroxyapatite and β-tri calcium phosphate microplasma spray coated pin intra-medullarly for bone repair in a rabbit model. *Ceramics International*, 8: 1377–1391.

17. Shukla B. P. 1989. A comparative evaluation of fresh autogenous vis-a-vis freeze dried and decalcified freeze dried segmental xenogenous bone grafts in dogs. MVSc dissertation, Deemed University Indian Veterinary Research Institute, Izatnagar.

18. Singh S. 1998. Reconstruction of segmental ulnar defect with tricalcium phosphate, calcium hydroxyapatite and calcium hydroxyapatite-bone matrix combination in rabbit. MVSc dissertation, Deemed University Indian Veterinary Research Institute, Izatnagar.

19. Trinder P. 1960. Calorimetric micro-determination of calcium in serum. *Analyst*, 85: 889–894.

20. Gomori G. 1942. A modification of calorimetric phosphorus determination. *Journal of Laboratory and Clinical Medicine*, 27: 955–976.

21. Kind P. R. and King E. J. 1954. Alkaline phosphatase activity determination in serum. *Journal of Clinical Pathology*, 7: 322–326.

22. Johnson K. A. and Watson A. D. J. 2000. Skeletal diseases. In *Textbook of veterinary internal medicine. Diseases of the dog and cat*, Ettinger S. J. and Feldman E. C. (Eds.). Philadelphia: W. B. Saunders, 1887–1916.

23. Rosol T. J. and Capen C. C. 1997. Calcium-regulating hormones and diseases of abnormal mineral (calcium, phosphorus, magnesium) metabolism. In *Clinical biochemistry of domestic animals*, Kaneko J. J., Harvey J. W., and Bruss M. L. (Eds.). San Diego: Academic Press, 619–702.

24. Meller Y., Kestenbaum R. S., Mozes M., Mozes G., Yagil R., and Shany S. 19984. Mineral and endocrine metabolism during fracture healing in dogs. *Clinical Orthopaedics and Related Research*, 187: 289–295.

25. Speed K. 1931. Blood serum calcium relation to the healing of fractures. *Journal of Bone and Joint Surgery*, 13: 58–67.

26. Pandey S. K. and Udupa K. N. 1981. Effect of anabolic hormone on certain metabolic response after fracture in dogs. *Indian Veterinary Journal*, 58: 37–41.

27. Saraswathy G., Sastry T. P., Pal S., Sreenu M., and Kumar R. V. S. 2004. A new bio-inorganic composite as bone grafting material: in vivo study. *Trends in Biomaterials and Artificial Organs*, 17: 37–42.

28. Suryawanshi S. B., Maiti S. K., Varshney V. P., and Singh G. R. 1999. Effect of anabolic hormones on haemato-biochemical changes in fracture healing in dogs. *Indian Journal of Animal Sciences*, 69: 404–406.

29. Nguyen H. Q., Deporter D. A., Pilliar R. M., Valiquette N., and Yakubovich R. 2004. The effect of sol-gel-formed calcium phosphate coatings on bone ingrowth and osteoconductivity of porous-surfaced Ti alloy implants. *Biomaterials*, 25: 865–876.

30. Areva S. and Jokinen M. 2008. Ensuring implant fixation and sol-gel derived ceramic coatings with special reference to TiO_2-based coatings. *Key Engineering Materials*, 377: 111–132.

31. de Groot K., Klein C. P. A. T., Wolke J. G. C., and de Blieck-Hogervorst J. M. A. 1990. Plasma-sprayed coatings of calcium phosphate. In *CRC handbook of bioactive ceramics*, Yamamuro T., Hench L. L., and Wilson J. (Eds.). Boca Raton, FL: CRC Press, 3–16.

32. Geesink R. G. T., de Groot K., and Christel P. A. K. T. 1987. Chemical implant fixation using hydroxylapatite coatings. *Clinical Orthopaedics and Related Research*, 225: 147–170.

33. Junior B. K., Beck T. J., Kappert H. F., Kappert C. C., and Masuko T. S. 1998. A study of different calcification areas in newly formed bone 8 weeks after insertion of dental implants in rabbit tibias. *Annals of Anatomy*, 180: 471–475.
34. Maiti S. K. and Singh G. R. 1995. Different types of bone grafts and ceramic implants in goats. A triple fluorochrome labeling study. *Indian Journal of Animal Sciences*, 65: 140–143.
35. Gibson C. J., Thornton V. F., and Brown W. A. B. 1978. Incorporation of tetracycline into impeded and unimpeded mandibular incisors of the mouse. *Calcified Tissue International*, 26: 29–31.
36. Nandi S. K., Ghosh S. K., Kundu B., De, D. K., and Basu D. 2008. Evaluation of new porous β-tri-calcium phosphate ceramic as bone substitute in goat model. *Small Ruminant Research*, 75: 144–153.
37. Nandi S. K., Kundu B., Ghosh S. K., De D. K., and Basu D. 2008. Efficacy of nano-hydroxyapatite prepared by an aqueous solution combustion technique in healing bone defects of goat. *Journal of Veterinary Science*, 9: 183–191.
38. Caulier H., van der Waerden J. P. C. M., Paquay Y. C. G. J., Wolke J. G. C., Kalk W., Naert I., and Jansen J. A. 1995. Effect of calcium phosphate (Ca-P) coatings on trabecular bone response: a histological study. *Journal of Biomedical Materials Research*, 29: 1061–1069.
39. Caulier H., van der Waerden J. P. C. M., Wolke J. G. C., Kalk W., Naert I., and Jansen J. A. 1997. A histological and histomorphometrical evaluation of the application of screw designed calcium phosphate (Ca-P) coated implants in the cancellous maxillary bone of the goat. *Journal of Biomedical Materials Research*, 35: 19–30.
40. Hulshoff J. E. G. and Jansen J. A. 1997. Initial interfacial healing events around calcium phosphate (Ca-P) coated oral implants. *Clinical Oral Implants Research*, 8: 393–400.
41. Hulshoff J. E. G., Hayakawa T., Van Dijk K., Leijdekkers-Govers A. F. M., Van der Waerden J. P. C. M., and Jansen J. A. 1997. A mechanical and histological evaluation of Ca-P plasma spray and magnetron-sputter coated implants in trabecular bone of the goat. *Journal of Biomedical Materials Research*, 36: 75–84.

11

Future Scope and Possibilities

In this chapter, the scope of future work is discussed. Here, we want to show the new possibilities of the microplasma spraying (MIPS) process in the biomedical field.

11.1 MIPS-HAp Coating on Complex and Contoured Implants

In the previous chapters, we showed that the microplasma spraying method has proved to be more efficient in terms of retaining a high degree of crystallinity after the spraying, phase purity, and possessing of high porosity level without compromising the macro- as well as micromechanical properties, e.g., nanohardness and Young's modulus. It has a unique microstructure to enhance the toughness and, to a relative extent, hinder the brittle crack propagation. It also shows no signature of large-scale delamination or rupture in a severe dynamic contact situation. In vitro followed by in vivo study established that the microplasma sprayed hydroxyapatite (HAp) coating also has the ability to heal bone defects in the animal model. Therefore, the MIPS-HAp coating has established its potential for future application to human beings for bone defect healing and also stable cementless fixation. Further, in comparison to macroplasma spraying (MAPS), the MIPS process can lead to achievement of better properties in terms of crystallinity, phase purity, porosity, and reduction of residual stress.

The research group at CSIR–Central Glass and Ceramic Research Institute (CGCRI), Kolkata, has already started the initiative for a human trial with the MIPS-HAp-coated metallic stems. Typical examples of HAp-coated complex-shaped metallic (such as SS316L, Ti-6Al-4V, etc.) implants coated by the microplasma spraying technique are shown in Figures 11.1 to 11.3 [1]. The MIPS technique was also able to coat the small implants for dental application (Figure 11.1), as well as the comparatively large complex-shaped stems in different size ranges for hip implantation (Figure 11.2) and its components, e.g., an acetabular shell (Figure 11.3).

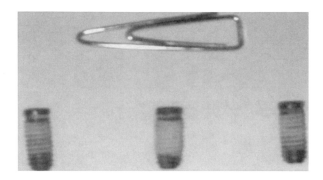

FIGURE 11.1 (See color insert.)
MIPS-HAp-coated Ti-6Al-4V dental implants (screw).

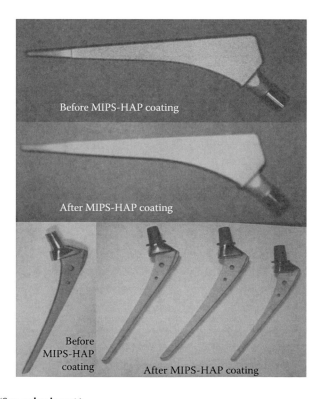

FIGURE 11.2 (See color insert.)
Before (grit blasted) and after MIPS-HAp coating on different-shaped SS316L hip implants
(length, 155 mm for largest size, 140 mm for smallest size).

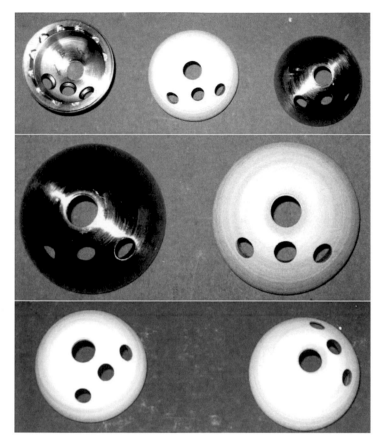

FIGURE 11.3 (See color insert.)
Before and after MIPS-HAp coating on SS316L acetabular shell (outer diameter, 48.35; inner diameter, 39.75 mm).

11.2 MIPS Coating of Other Calcium Phosphates (TCP, BCP, etc.)

The other calcium phosphates, such as tricalcium phosphate (TCP), biphasic calcium phosphate (BCP), etc., could also be coated on metallic implants for patient and problem-specific application purposes. The surface reactivity of TCP is the highest, followed by that of BCP and then HAp. Therefore, if any specific medical case demands fast resorption, then coating of TCP would be the best choice, and for comparatively controlled resorption, BCP coating would be used. Further, the MIPS-TCP coating also showed promising properties in both in vitro and in vivo trials. This was recently also reported from our group [2]. However, more research should be carried out for the

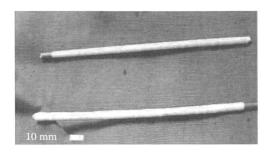

FIGURE 11.4 (See color insert.)
MIPS-TCP-coated intramedullary pin. (Reprinted/modified from Dey et al., *Ceramics International*, 8: 1377–1391, 2011. With permission from Elsevier.)

systematic micromechanical as well as macromechanical characterization. A typical example of a MIPS-TCP-coated intramedullary pin is shown in Figure 11.4 [2].

11.3 MIPS-HAp Coatings on C/C Composites

The Young's moduli of the metallic implants are always higher than that of the bone, which can often cause the stress shielding effect [3]. Moreover, the high density of metallic implants ultimately produces a heavy load on the parent bone. C/C composites having values of modulus, hardness, and density very near to those of bone are a promising replacement material for metallic implants. Research on the MIPS-HAp coating on the C/C composite would be another promising subject for future possibility. HAp coatings on C/C composite substrates were already developed by Sui et al. [4, 5] and Xin-bo et al. [6] by both MAPS and electrochemical deposition methods. The MAPS-HAp showed superior shear strength values. For example, MAPS-HAp coating deposited with higher power (45 vs. 25 kW) showed much higher shear strength values (e.g., 7.4 vs. 2.1 MPa). HAp coatings on C/C composites were also developed by the electrochemical deposition method [6]. However, a nanoindentation study of HAp coatings on C/C has yet to be explored. The present authors [1] developed MIPS-HAp on C/C composite substrate. The microstructure of a polished cross section of MIPS-HAp coating on the C/C composite substrate is shown in Figure 11.5. The initial data on MIPS-HAp coating on C/C composites also showed good micromechanical properties. The nanohardness and the reduced modulus were found as 6.15 ± 1.72 GPa and ~130.2 ±1 8.1 GPa, respectively, from the nanoindentation test at 1000 µN load. More research work should be carried out to prove its efficacy in both in vitro and in vivo trials.

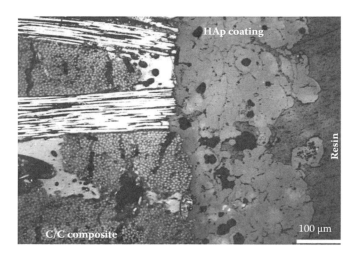

FIGURE 11.5
Microstructure of a polished cross section of the MIPS-HAp coating on the C/C composite substrate.

11.4 Second Phase Incorporation in HAp Coatings

To improve the mechanical properties, in particular fracture toughness, researchers have incorporated second phases with HAp, e.g., yttria-stabilized zirconia (YSZ) [7–9], titanium dioxide [10, 11], carbon nanotube (CNT) [12], and fluorine [13].

Incorporation of CNT in HAp deposited by the MAPS method showed improvement of fracture toughness from 0.39 $MPa.m^{0.5}$ for pure HAp to 0.61 $MPa.m^{0.5}$ for the HAp-CNT composites [12], along with a concomitant improvement in crystallinity from ~54% for pure HAp to ~80% for the HAp-CNT composites. However, even without incorporation of a second phase, the pure MIPS-HAp coating showed a higher toughness of 0.6 $MPa.m^{0.5}$ (as mentioned in Chapter 7) and a crystallinity of ~81% that could be raised to as high as ~92% following post-deposition heat treatment (as mentioned in Chapter 4). However, the most important point was that in cell culture studies with human osteoblast hFOB 1.19 cells, the HAp-CNT composite coatings showed a satisfactory result [12].

Fu and his coauthors [7–9] worked on deposited by the MAPS technique to improve not only toughness but also hardness, HAp-YSZ composite's modulus, and bonding strength properties of HAp-based coatings. The toughness improved from 0.55 $MPa.m^{0.5}$ to about 1.1 $MPa.m^{0.5}$, along with some improvement of both hardness and bonding strength for the HAp-YSZ composites [9]. It was further reported by the same group that enhancing the amount of YSZ in the HAp-YSZ composite, the toughness could even be hiked to 1.5 MPa.m$^{0.5}$ [7]. The toughness of these HAp-YSZ composites appeared to also be sensitive to the process used to mix the starting YSZ and HAp powders [8].

Li et al. [10, 11] reported that the incorporation of TiO$_2$ in HAp by the high-velocity oxy fuel (HVOF) technique can also enhance the fracture toughness and shear strength properties. The toughness value was increased from 0.48 to 1.67 MPa.m$^{0.5}$ as TiO$_2$ was incorporated in HAp [10]. However, their data further suggest that a trade-off point exists between enhancements in toughness, Young's modulus, and bonding strength, on the one hand, and the amount of TiO$_2$ that can be added as reinforcement, on the other.

It has been reported for sol-gel dip-coated fluorine-doped HAp coating [13] that both fracture toughness (0.31 MPa.m$^{0.5}$) and bonding strength (27 MPa) properties are enhanced with respect to those (0.12 MPa.m$^{0.5}$ and 19 MPa) measured for the undoped or pure HAp coatings. The development of MIPS composite HAp coatings has yet to be explored, and hence may provide ample scope of future research and development in this area.

11.5 Nanostructured Plasma Sprayed HAp Coating

It is very difficult to deposit nanostructured powders directly by any thermal spraying method, as the nanostructured powders have a tendency to agglomerate. In the plasma spraying or thermal spraying technique, free-flowing powder is required. But as it agglomerates, it is almost an impossible task to make plasma spraying with a nanostructured powder. To overcome this limitation, the starting nanostructured HAp particles were processed as micron-size granules, and thus it was possible to make plasma spraying of these granules, as they could flow freely. Figure 11.6a showed the typical microstructure of spray-dried spherical-shaped HAp powder of several microns in size, while the higher-magnification field emission scanning electron microscopy (FESEM) image presented in Figure 11.6b showed how the nanosize particles consolidated and ultimately form a micron-size granule, shown in the lower-magnification image presented in Figure 11.6a. Development of MAPS sprayed HAp coatings from nanostructured HAp granules was reported by Renghin et al. [14] and Han et al. [15]. Further, Zakeri et al. [16] reported development of MAPS HAp-TiO$_2$ composite coatings, starting from the composite form of powder granule of nanostructured HAp-TiO$_2$. Further, utilizing liquid precursor, deposition of HAp by the liquid precursor plasma spraying (LPPS) process was also reported by Huang and his coworkers [17, 18]. In fact, the LPPS process demonstrated excellent control on the coating microstructure by simply adjusting the solid content

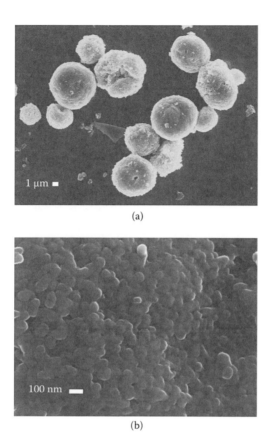

(a)

(b)

FIGURE 11.6
Spray-dried nanostructured HAp powder in (a) granule form and (b) higher magnification showing nanostructured HAp particles.

of the HAp in the liquid precursor. The beauty of the LPPS process lies in that both almost fully dense and highly porous HAp coatings can be obtained by it. However, there is no report yet of an attempt where the MIPS is married to the LPPS process; i.e., a low plasmatron power, as used in MIPS, is utilized for the LPPS process. Therefore, this can be another area of prospective future research with the possibility of a high payoff, if successful. Another prospective area is MIPS of granules with nanostructured HAp particle or nanostructured HAp second phase composites. If successful, this technique can take advantage of the greenness of the MIPS process, on the one hand, and the nanocomposite design philosophy, on the other, thereby providing the future possibility for development of harder, stronger, tougher, and more reliable bioactive ceramic coatings by the MIPS process.

Finally, conclusions of this book are documented in Chapter 12.

References

1. Dey A. 2011. Physico-chemical and mechanical characterization of bioactive ceramic coating. PhD dissertation, Indian Institute of Engineering Technology (formerly Bengal Engineering and Science University), Shibpur, Howrah, India.
2. Dey A., Nandi S. K., Kundu B., Kumar C., Mukherjee P., Roy S., Mukhopadhyay A. K., Sinha M. K., and Basu D. 2011. Evaluation of hydroxyapatite and β-tri calcium phosphate microplasma spray coated pin intra-medullarly for bone repair in a rabbit model. *Ceramics International*, 8: 1377–1391.
3. Ratner B. D., Hoffman A. S., Schoen F. J., and Lemons J. E. 2004. *Biomaterials science: an introduction to materials in medicine*. Elsevier Academic Press.
4. Sui J.-L., Li M.-S., Lu Y.-P., Yin L.-W., and Song Y.-J. 2004. Plasma-sprayed hydroxyapatite coatings on carbon/carbon composites. *Surface and Coatings Technology*, 176: 188–192.
5. Sui J.-L., Li M.-S., Lu Y.-P., and Bai Y.-Q. 2005. The effect of plasma spraying power on the structure and mechanical properties of hydroxyapatite deposited onto carbon/carbon composites. *Surface and Coatings Technology*, 190: 287–292.
6. Xiong X.-B., Zeng X.-R., Zou C.-L, Li P., and Fan Y.-B. 2009. Influence of hydrothermal temperature on hydroxyapatite coating transformed from monetite on HT-C/C composites by induction heating method. *Surface and Coatings Technology*, 204: 115–119.
7. Fu L., Khor K. A., and Lim J. P. 2002. Effects of yttria-stabilized zirconia on plasma-sprayed hydroxyapatite/yttria-stabilized zirconia composite coatings. *Journal of the American Ceramic Society*, 85: 800–806.
8. Fu L., Khor K. A., and Lim J. P. 2001. Processing, microstructure and mechanical properties of yttria stabilized zirconia reinforced hydroxyapatite coatings. *Materials Science and Engineering A*, 316: 46–51.
9. Fu L., Khor K. A., and Lim J. P. 2000. Yttria stabilized zirconia reinforced hydroxyapatite coatings. *Surface and Coatings Technology*, 127: 66–75.
10. Li H., Khor K. A., and Cheang P. 2002. Titanium dioxide reinforced hydroxyapatite coatings deposited by high velocity oxy-fuel (HVOF) spray. *Biomaterials*, 23: 85–91.
11. Li, H., Khor K. A., and Cheang P. 2002. Young's modulus and fracture toughness determination of high velocity oxy-fuel-sprayed bioceramic coatings. *Surface and Coatings Technology*, 155: 21–32.
12. Balani K., Anderson R., Laha T., Andara M., Tercero J., Crumpler E., and Agarwal A. 2007. Plasma-sprayed carbon nanotube reinforced hydroxyapatite coatings and their interaction with human osteoblasts in vitro. *Biomaterials*, 28: 618–624.
13. Zhang S., Wang Y. S., Zeng X. T., Khor K. A., Weng W., and Sun D. E. 2008. Evaluation of adhesion strength and toughness of fluoridated hydroxyapatite coatings. *Thin Solid Films*, 516: 5162–5167.
14. Renghinia C., Girardin E., Fomin A. S., Manescu A., Sabbioni A., Barinov S. M., Komlev V. S., Albertini G., and Fiori F. 2008. Plasma sprayed hydroxyapatite coatings from nanostructured granules. *Materials Science and Engineering B*, 152: 86–90.

15. Han Y., Xu K., Montay G., Fu T., and Lu J. 2002. Evaluation of nanostructured carbonated hydroxyapatite coatings formed by a hybrid process of plasma spraying and hydrothermal synthesis. *Journal of Biomedical Materials Research,* 60: 511–516.
16. Zakeri M., Hasani E., and Tamizifar M. 2013. Mechanical properties of TiO_2-hydroxyapatite nanostructured coatings on Ti-6Al-4V substrates by APS method. *International Journal of Minerals, Metallurgy and Materials,* 20: 397–402
17. Huang Y., Song L., Liu X., Xiao Y., Wu Y., Chen J., Wu F., and Gu Z. 2010. Hydroxyapatite coatings deposited by liquid precursor plasma spraying: controlled dense and porous microstructures and osteoblastic cell responses. *Biofabrication,* 2: 045003.
18. Huang Y., Song L., Liu X., Xiao Y., Wu Y., Chen J., Wu F., and Gu Z. 2010. Characterization and formation mechanism of nano-structured hydroxyapatite coatings deposited by the liquid precursor plasma spraying process. *Biomedical Materials,* 5: 054113.

Printed and bound by CPI Group (UK) Ltd, Croydon, CR0 4YY

22/10/2024

01777647-0001

12

Conclusions

The major emphasis of Chapter 1 was put on two aspects: the different types of (1) biomaterials and (2) bioactive materials. In the process, we also highlighted what exactly is meant by biocompatibility, and why we need biomaterials to be biocompatible. Gradually, the discussion zeroed in to focus on bioactive ceramics, and especially on the bioactive ceramic material hydroxyapatite (HAp). Finally, we emphasized its usage as a bioactive ceramic coating. In addition, a glance was made at natural biomaterials like bone and teeth. It was shown, in a nutshell, how both of these natural biomaterials possess hierarchically designed functionally graded microstructures that not only span from a nanometer to a meter order of dimensions, but also help to perform multiple, repetitive, simultaneous, and predefined actions in a smart and intelligent manner. The idea of such a natural nanobiohybrid composite was finally shown to transgress into both the need and challenges involved in the development of surface-engineered biometallic implants.

Chapter 2 was dedicated to describe the basics of the plasma spraying process. It was a major focus to elucidate what is the macroplasma spraying process and what is the microplasma spraying process. The role of the plasmatron power on the development of coating was identified in connection with the illustration of how a coating formation physically takes place on a given substrate, and how the growth of the coating along the thickness direction happens by this process. The next point of focal attention was to describe what the relative advantages of the microplasma spraying (MIPS) process over the macroplasma spraying (MAPS) process are, and what the geneses of such advantages are. The major part of the following discussion in this chapter was naturally dedicated to two most important aspects: (1) the uniqueness of the MIPS process as a coating manufacturing technique and (2) its application. Additional, yet very important, information that was offered in this chapter was a brief, yet as far as possible complete, picture of the other, more well-known coating techniques, and a comparative narration of their relative merits and demerits.

Chapter 3 had a specific focus on HAp coatings and their applications. At first, the plasma spraying parameters of interest were identified. Next, a detailed discussion was presented on the influence of plasma spraying parameters on the characteristics of HAp coatings. To the researcher and the manufacturer, a major concern of practical interest is which coating process, e.g., a low-temperature-based process or a high-temperature-based process, is to be chosen for a given end application. Therefore, the relative merits and demerits of high-temperature and low-temperature-assisted HAp coating deposition processes were critically analyzed. However, it is well known that in any given field, knowledge of the most current state-of-the-art situations is an absolute must to proceed further. Hence, the state-of-the-art scenarios of the nanostructured HAp coatings, HAp composite coatings, and doped HAp coatings were described, with an analysis of their relative limitations and advantages. Finally, with a view to highlight the major success of MIPS-HAp coatings developed in recent times, the chapter provided the most focused yet elaborate attention on the relative efficacies of the HAp coatings deposited by the MAPS and MIPS processes.

Now it needed to be emphasized that the microstructural elements that make a coating up are a matter of extraordinary importance. This is what Chapter 4 was all about. There we discussed not only the porosity, crystallinity, and stoichiometry, but also the splat size and its distribution, microcrack size and its distribution, etc., of the HAp coatings. Although situations pertaining to both MAPS- and MIPS-HAp coatings were discussed, the truth is that on a comparative scale, relatively more detailed discussion was devoted to the latter type of coatings rather than the former. It was described in detail how a phase-pure MIPS-HAp coating with 80–92% crystallinity and a relatively higher porosity of ~20% in the plan and ~11% across the cross section is developed from HAp granules of appropriate size fraction. In addition, it was explained why the degree of crystallinity and the volume percentage porosity of the MIPS coatings were much higher than those usually reported for the conventional MAPS-HAp coatings. Post-heat treatment improved the degree of crystallization in the coating. Interesting roles of the dehydroxylation and rehydroxylation processes in the as-sprayed and post-heat-treated coatings were established from the analysis of the relevant Fourier transform infrared (FTIR) data. Scanning electron microscopy (SEM) and field emission scanning electron microscopy (FESEM) studies of the coating confirmed that the coating microstructure was extremely heterogeneous. They revealed that the MIPS-HAp coatings were dotted with the presence of macro- and microcracks, inter- and intrasplat cracks, macropores, micropores, and unmelted HAp particles. An image analysis technique provided information on various important microstructural parameters, e.g., average splat size, splat aspect ratio, and micropore and microcrack sizes. The coating had a splat size of ~50–70 μm, macropore size of ~10–50 μm, and micropore size of ~1 μm. The thickness of the MIPS-HAp coating was measured as ~210 μm. As revealed by FESEM photomicrographs and volume percent porosity data obtained from image analysis of the heterogeneous microstructure, the anisotropy present in the nanohardness data was linked to the larger volume percent porosity (p), as well as higher spatial density of planar defects, pores, and cracks in the plan section over those in the cross section. In this connection, a qualitative model was schematically developed to pictorially depict the genesis of anisotropy in nanohardness of the present MIPS-HAp coating. Finally, the best fit for the porosity dependence of combined hardness data from literature and the present work was given by an empirical generic equation of the form $X = X_0 \exp(-bp)$, where X stands for nanohardness (H) or Young's modulus (E), as the case may be. The exponential dependence of E and H on porosity estimated E_0 as 117.4 GPa and H_0 as 5.92 GPa, which were comparable to literature data, where the suffix 0 stands for theoretically dense, e.g., zero-porosity, material. Thus, the volume fraction open porosity also played an important role in anisotropy of nanohardness in the present MIPS-HAp coating. Further, it was elucidated why the Young's modulus of thermal spray deposits is influenced not only by the volume percent of open porosity, but also by the pore morphology. The shape of the pore or void could be spherical, elliptical, or a superimposition of spherical and elliptical pores/voids altering the aspect ratio. Experimental evidence gathered for the present MIPS-HAp coating also suggested that the pores could be spherical, elliptical, penny, or thin crack shaped in nature. Therefore, taking this aspect of the microstructure into account, the Young's moduli along the plan section, E_{11}, and cross section, E_{22}, of the coating had been predicted by the spherical pore model, elliptical pore model, penny-shaped pore model, and thin crack-shaped pore model. Finally, it was illustrated that values of E_{11}, E_{22} and the anisotropy factor, i.e., the ratio between E_{22} and E_{11}, predicted with the penny-shaped pore model matched the best among all models with the experimentally measured data of the present MIPS-HAp coatings.

It is important not only to develop a MIPS-HAp coating with good nano-mechanical properties, but also to make sure that they are stable when in contact with the physiological medium during the service lifetime. Thus, the subject matter of Chapter 5 was naturally chosen to be the typical results of a systematic investigation on the dissolution of MAPS- and MIPS-HAp coatings following immersion in the simulated body fluid solution. As the amount of studies reported on the latter type of coatings is far from significant, major importance was devoted to the in-depth investigation of the dissolution of MIPS-HAp coatings, rather than on the MAPS-HAp coatings. The first emphasis was placed on the different types of physical, chemical, and microstructural changes that can happen following immersion. Simultaneously, it was attempted to illustrate the signatures corresponding to these changes and the relevant experimental techniques that are to be utilized to pick these signatures up. Finally, it was attempted to explain how and why these changes happen. The data obtained from the inductively coupled plasma atomic emission spectroscopy (ICP-AES) results confirmed that after the first day of immersion in the simulated body fluid (SBF) solution the dissolution of Ca and P was dominant. The additional results found from X-ray diffraction (XRD), FTIR, SEM, environmental scanning electron microscopy (E-SEM), and energy-dispersive X-ray spectroscopy (EDX) experiments, however, confirmed that after the fourth day and up to 14 days after immersion, the deposition of fine-needle-like nanostructured apatites was the dominant process. The results obtained from the ICP-AES experiments confirmed that when the MIPS-HAp coatings deposited on biomedical-grade

SS316L substrates were immersed in the SBF solution up to a time period of as high as 2 weeks of immersion time, the leaching of toxic chromium and iron metal ions did not happen. The results presented in Chapter 5 strongly suggest that the stability of the MIPS-HAp-coated implants on SS316L for biomedical prosthesis applications is reliable.

As described in the aforesaid chapters, the MIPS-HAp coatings have good nanomechanical properties and are stable in SBF solution. But having these attributes cannot necessarily guarantee that they will have macromechanical properties like bonding strength, shear strength, fatigue properties, etc., acceptable or good enough for practical applications. That is why in Chapter 6 we discussed why macromechanical properties like the bonding strength, shear strength, fatigue properties, etc., of HAp coatings are particularly important when HAp coatings will be used in in vivo implantation. For HAp coatings processed by different methods, the comparison of bonding strengths and other properties was provided. Of course, comparatively more importance was devoted to plasma sprayed coatings in general, and the MIPS-HAp coatings in particular. The different factors influencing bonding strength, shear strength, and fatigue behavior were discussed in detail. Further, under the controlled prefixed stress level, how HAp-coated metallic substrate will behave during the three-point bending test was discussed, with particular emphasis on propagation of the crack front during the test. The results of the crack front propagation showed that the microstructure of the MIPS-HAp coating had an inherent strain-tolerant nature.

Now the question that naturally arises is: How good are the mechanical properties of the MIPS-HAp coatings at the microstructural length scale? That is why the major effort in Chapter 7 was to understand the different aspects of the micro/nanomechanical properties, like nanohardness, elastic modulus, and fracture toughness of various HAp coatings evaluated at the local microstructural length scale. In the discussion, more importance was given to the relevant micro/nanomechanical properties of plasma sprayed HAp coatings in general, and MIPS-HAp coatings in particular. This particular chapter discussed in detail the nanoindentation technique. It was emphasized that this technique is one of the potential tools to characterize micro/nanomechanical properties at the local microstructural length scale of thin films and coatings. The literature scenarios regarding nanohardness, elastic modulus, and fracture toughness of HAp coatings were explicitly explained. Scatter of data that had happened during the nanoindentation study on MIPS-HAp coatings was discussed. Further, it was illustrated how this scatter can be treated in terms of a two-parameter Weibull distribution. MIPS-HAp coatings contain defects, microcracks, and porosity that finally affect the indentation size effect. In this context, the inherent complexity of the indentation size effect was explained. In addition, the huge importance of fracture toughness evaluation at the microstructural length scale was highlighted. More stress was put on indentation-based fracture toughness measurement in general, and nanoindentation-based fracture toughness measurement in particular. Finally, the scenario of microstructure/crack interaction was shown to explain why MIPS-HAp coatings have high fracture toughness. The results obtained in Chapter 7 finally highlighted that the MIPS-HAp coatings had nanohardness, Young's modulus, and fracture toughness comparatively higher than those of the HAp coatings deposited by the MAPS process.

So far the MIPS-HAp coatings have shown both good nano- and macromechanical properties under quasi-static conditions and good stability in SBF. But that does not necessesarily ensure that they will not fail during exposure to dynamic load when in service. Thus, an essential requirement for further application of the MIPS-HAp coatings is the characterization and qualification in terms of micro/nanotribological properties. Hence, the focus of Chapter 8 was solely on the tribological properties, investigated by both scratch and fatigue tests of HAp coatings. Nanoscratch tests on the MIPS-HAp coatings, with both constant and ramping loads, showed no signature of microfracture, delamination, or peel-off, thus possibly corroborating the presence of a strain-tolerant microstructure. The microscratch test on the MIPS-HAp coatings at the ramping normal loads also did not show any large-scale delamination or coating peel-off, except at very high loads, e.g., 10–10.6 N. The average coefficient of friction was about ~0.4 for as-sprayed, but reduced further for the polished MIPS-HAp coating. In general, μ of the as-sprayed MIPS-HAp coating was slightly higher than that of the polished MIPS-HAp coating, due to the higher surface roughness (R_a ~ 0.4 μm) and porosity (~20%) of the former. The coefficients of friction for the MIPS-HAp coating were SBF immersion time dependent. However, the microscratch tests conducted at even a relatively higher ramping load (e.g., 10–10.6 N) showed no large-scale delamination or coating peel-off, which proved the stability of the coating after immersion in a synthetically produced body fluid environment. A thorough study of the relevant literature revealed that there has been no fatigue study reported yet for MIPS-HAp coatings. Thus, the MIPS-HAp coatings appeared to have not only good macro- and nanomechanical properties, but also good micro- and nanotribological properties. Also, these MIPS-HAp coatings had good stability in SBF solution, as discussed in the earlier chapters.

But it needs to be emphasized here that even when a given coating has good mechanaical and tribological properties and compositional stability, it could still fail from the presence of residual stress. Hence, this particular aspect of utmost importance was highlighted for HAp coatings in Chapter 9. The origin of residual stress and why it is harmful for in vivo application of HAp coatings were explained in this chapter. Further, several techniques for measuring residual stress were discussed. In addition, the relative merits and demerits of different methods for measurement of residual stress were critically assessed. The residual stresses of HAp coatings prepared by MAPS, MIPS, and other processes were extensively compiled. Further, the efficacy and applicability of the newly evolved nanoindentation-based technique was established by examining the residual stress state in MIPS-HAp coatings deposited on Ti-6Al-4V and SS316L substrates. It was further demonstrated that the magnitude and nature of residual stress states estimated by the nanoindentation-based measurement technique closely matched those evaluated by the well-established XRD-based measurement technique. The results presented Chapter 9 confirmed that the MIPS-HAp coatings had compressive residual stress active at the surface.

Thus, the results presented in Chapters 4–9 established the reasonably good combination of macro- and nanomechanical properties, micro- and nanotribological properties, high fracture toughness, and a strain-tolerant microstructure with high porosity and crystallinity, along with the presence of a small compressive residual stress at the surface of the MIPS-HAp coatings. They also possessed good stability even when exposed to SBF solutions for a prolonged period of up to 2 weeks. Therefore, it was decided to examine the efficacy of the MIPS-HAps coatings in a given animal model. That is why Chapter 10 presented, for the very first time, a study on the comparative healing performance efficacies of uncoated and MIPS-pure HAp-coated intramedullary SS316L pins for bone defect fixation in New Zealand white rabbits. The results of the biochemical, histological, radiographic, and fluorochrome labeling studies proved more bone bonding and osseointegration in the MIPS-HAp-coated metallic pins. In fact, such a generic process has led to complete bone defect healing that was not possible to attain with the uncoated metallic pins. In other words, the MIPS-HAp-coated intramedullary SS316L pins have a performance efficacy that is much superior to that of uncoated metallic pins for repairing bone defects in the New Zealand white rabbits. In addition, it has been well documented that the superior efficacy of the MIPS-HAp-coated metallic implants matched the results of additional research conducted by other researchers in goat and dog models. Thus, it may be suggested that MIPS-HAp-coated or CaP-coated metallic implants will surely have a very promising future for repairing bone defects. The results derived from these developmental efforts strongly suggest that with further optimization of the microstructure, mechanical, functional, and tribological properties, on the one hand, and the residual stress, on the other, a possibility for human trials of the MIPS-HAp coatings may emerge in days to come.

There is no doubt that even given the phenomenal development in many varieties of bioactive ceramic coatings, especially the MIPS-HAp coatings, the scope of further enhancement in their properties and efficacy is tremendous. Hence, Chapter 11 briefly discussed the scope of future work in this exciting field of research. Many new possibilities of the MIPS process in biomedical applications have been hinted at to encourage more developmental and application-oriented research in this area, where every small success can indeed mean a big stride in relief for millions in the ailing population in both developing and developed nations. Finally, all major conclusions were summarized in this chapter.

Index